EYES IN THE SKY

EYES IN THE SKY

The Secret Rise of GORGON STARE and How It Will Watch Us All

ARTHUR HOLLAND MICHEL

Houghton Mifflin Harcourt
Boston   New York
2019

For information about permission to reproduce selections from this book, write to trade.permissions@hmhco.com or to Permissions, Houghton Mifflin Harcourt Publishing Company, 3 Park Avenue, 19th Floor, New York, New York 10016.

hmhbooks.com

Library of Congress Cataloging-in-Publication Data
Names: Michel, Arthur Holland, author.
Title: Eyes in the sky : the secret rise of Gorgon Stare and how it will watch us all / Arthur Holland Michel.
Description: Boston : Houghton Mifflin Harcourt, 2019. |
Includes bibliographical references and index.
Identifiers: LCCN 2018046435 (print) | LCCN 2018048027 (ebook) |
ISBN 9780544971660 (ebook) | ISBN 9780544972001 (hardcover)
Subjects: LCSH: Intelligence service — United States. | Aerial surveillance — United States. | Electronic surveillance — United States. | Civil rights — United States. | Privacy, Right of — United States.
Classification: LCC JK468.I6 (ebook) | LCC JK468.I6 M5155 2019 (print) |
DDC 363.2/32 — dc23
LC record available at https://lccn.loc.gov/2018046435

Book design by Chrissy Kurpeski
Typeset in Warnock Pro

Printed in the United States of America
DOC 10 9 8 7 6 5 4 3 2 1

For my sister, Gaby; my brothers — Simon, Maxi, Quino;
my pa, Andreas; my ma, Rachel;
and Erin . . .

There Earth produc'd the Gorgon, dreadful monster.

— EURIPIDES, *Ion*

Contents

PART III:

ANOTHER LONG ROAD

Introduction

On a clear day in the spring of 1862, two orbs rose quietly into the skies above southeastern Virginia. For the soldiers in the Confederate encampment below, it must have been a wondrous, if demoralizing, sight. These strange forms were Union hot-air balloons, each with two observers in the basket, spies in the firmament. If the observers had cameras — and word was that they did — they would be able to produce perfect records of what they saw, instant maps of the Confederate positions that would soon, no doubt, be conveyed to the upper reaches of the Union Army's chain of command.

"We watched with anxious eyes their beautiful observations as they floated high up in the air," Confederate general James Longstreet later wrote of the encounter. To Longstreet and his troops, it was clear they were witnessing a turning point in the history of war. It was also clear that there was nothing they could do about it. The balloons, Longstreet lamented, were "well out of range of our guns."

One hundred and fifty-five years later, on a searingly hot afternoon in June 2017, Steve Suddarth, a former US Air Force colonel, is radioing air traffic control for clearance to take off from Albuquerque International Sunport, the city's main airport. The flight, he tells the tower, is a "photo mission." Riding shotgun in Suddarth's white single-prop Cessna, I am not sure I would have used such a mild term to describe our plan. The plane is equipped with a military-grade surveillance camera, and we intend to watch wide tracts of the city in a way many of its residents probably never imagined was even possible.

Air traffic control grants us free rein within a broad swath of airspace 12,000 feet directly above the city center. As we climb out of the airport through the choppy desert air, the landscape falls quickly from our wings. Soon enough, the entire town lies rolled out below us, shimmering in the afternoon sun.

Mounted on the windshield is a tablet displaying a satellite map of the city. Using a wireless keyboard he keeps stowed beside his seat, Suddarth clicks on a large white building in the center of town and a new image appears on the screen. It looks like a satellite picture, superimposed perfectly over the first, showing all the same buildings and roads. But when Suddarth zooms in, the image becomes like a sample of pond water seen through a microscope — full of life. There are cars, trucks, and buses on the streets, stopping and starting at traffic lights, turning at junctions, parking along sidewalks. Everything is moving. This is live video delivered from the camera. The frame covers two-dozen city blocks.

Suddarth explains that the camera is linked to the Cessna's autopilot, which is programmed to steer the airplane in such a way that the target area never falls out of the frame. To demonstrate, Suddarth takes his hands off the controls, and without pause the airplane pitches left. Once the airplane finds itself in range, he says, the camera will put us into a wide circular orbit. Our particular target — a mall, Suddarth tells me cheerfully — remains dead center on the screen.

After completing the loop, we swing the camera over to the University of New Mexico. On the athletics field, tiny shadows dart about on the screen: students exercising.

From the university, we turn to Sandia Heights, an affluent neighborhood on the northeastern edge of the city. Using the tablet, I zoom in on a four-lane road cutting across the area. A bright blue car is turning onto a tree-lined street. I follow it. Suddarth is talking about the system's technical details, but I am absorbed in the story unfolding before me. The car is driving slowly, and it seems to be meandering. Maybe they're lost, I think, or perhaps they're on a joyride. *Or maybe they're up to something.*

And then, sitting in this odd little plane steered by a robotic camera, I start to feel uncomfortable. The people in the car don't know they are being watched. Even if they did, like Longstreet and his soldiers there is not much they could do about it. More to the point, I am troubled because, whatever their story, it is none of my business.

The story of how we got here, though, *is* my business, and theirs too. Even before 1862 — before, even, the advent of manned flight — it was obvious that an eye in the sky would be a great military asset. As seen from above, your adversary is an open book. You can determine where they're hiding, track them wherever they go, even anticipate their next move. By the First World War, every major military power had eyes in the sky, with dramatic consequences for those on the ground. Battles were won and lost on the strength of aerial surveillance. One reporter, writing about the airborne camera in 1917, declared it to be "many times deadlier than its equivalent weight of high explosive." In the years since, militaries never stopped seeking a wider, sharper, more penetrating view of the ground.

I started studying airborne spycraft in 2012, when I was a senior at Bard College, a small liberal arts school nestled on the banks of the Hudson River about two hours north of New York City. Looking for a new academic diversion beyond my regular coursework, I had founded, along with my freshman-year roommate Dan Gettinger, the Center for the Study of the Drone, a research initiative that sought to put Bard's protean approach to intellectual inquiry to the many vexing issues posed by the advent of unmanned vehicles. One of our earliest projects involved comparing civilian accounts of military drone attacks to the descriptions of Harpies in classical literature, namely the *Argonautika*, a third-century epic poem by Apollonius of Rhodes.

Our timing, it turned out, had been providential; by the time Dan and I graduated, public interest in drones had swelled so much that we figured we might still get some more mileage out of our idea. The day after we donned the cap and gown, I began working for the college as the codirector of the center.

Over the years that followed in that role, I studied many disquieting technologies — Hellfire missiles, laser-guided bombs, the prospect of Amazon delivery drones — but there was only one that reliably haunted my dreams.

It's called WAMI (pronounced *whammy*), which stands for wide-area motion imagery. This is what Steve Suddarth and I used to spy on the unsuspecting residents of Sandia Heights. If the aerial-surveillance technology of 1917 was more dangerous than TNT, this is a weapon of mass destruction. By name and by design, WAMI watches a very broad area, in some cases even a whole city. It is so powerful that when you want to look at something closely, you simply zoom in on the image itself; the camera continues to record the entire view. While I tracked the blue car, the camera was still watching the rest of the neighborhood, recording residents' every move.

In this way, WAMI creates unprecedented new forms of power for those who control it and equally unprecedented woes for those on the ground who find themselves in its crosshairs. It is emblematic of the path of all forms of modern surveillance, toward a view of the surveilled that is more penetrating, more contiguous, more detailed, and entirely inescapable. It has other names: WAPSS (for wide-area persistent surveillance system), WFOV (wide field of view), WAAS (wide-area airborne surveillance). I call it the all-seeing eye.

The idea of the all-seeing eye originated with a small, determined group of engineers working against the clock to solve an urgent crisis arising from the military operations in Iraq and Afghanistan. They designed the technology around a simple logic: the more ground you can cover, the more people you can watch. And the more people you watch, the better your chance of catching the "bad guys." It is a line of thinking not all that dissimilar from the National Security Administration's "collect it all" approach to surveillance, which originated during the same wars. In the NSA's case, it was a philosophy that begat, with the Snowden leaks in 2013, one of the greatest privacy scandals in modern history.

In the case of WAMI, it gave rise to a formidable weapon that appears to have saved hundreds, if not thousands, of lives on the battlefield — by enabling the killing of countless bad guys. As a result of these exploits, it has secured a place in the penumbral pantheon of American spy tools and inspired many others to pursue a wide-area view of their own.

Those pursuing WAMI technology today include a number of groups much closer to home. One surveillance-company executive put it to me this way: even a camera one-tenth the size of the Air

Force's Gorgon Stare—which plays a starring role in this book, if such a thing can be said of a camera—is still thousands of times more powerful than that employed on regular police and FBI aircraft. How could any law enforcement agency *not* desire such a technology?

For this reason, there are mounting efforts to bring the all-seeing eye to US law enforcement, part of a drastic expansion of aerial surveillance of all kinds in our skies. A particularly important test case took place in the city of Baltimore, where a determined entrepreneur has been striving to demonstrate that WAMI will not only crack otherwise unsolvable crimes—in particular, shootings and assaults—but also deter people from engaging in criminal activity in the first place.

Such cases point to a nearly certain prognosis: someday, most major developed cities in the world will live under the unblinking gaze of some form of wide-area surveillance. Those that don't will pick up related technologies—often also military in origin—based on the same maximalist philosophy of surveillance and control. It matters not that WAMI was first fielded, like so many of these other weapons, for a circumscribed purpose. Any surveillance technology, if it is powerful enough, will always take on a life of its own. Once it's out, the all-seeing cat never goes back in the bag.

Not only that—it evolves. Labs, intelligence agencies, and private firms in America and abroad are working to develop the next generation of surveillance technologies, designed to watch the broadest possible area at the highest possible resolution. As a result, wide-area surveillance is getting cheaper, lighter, faster, and more powerful.

And smarter, too. In their pursuit of a wider view, the engineers and intelligence agencies behind WAMI created a device so powerful, humans could no longer keep up. When I was flying over Albuquerque, I could track only one car at a time, though the camera was recording an area encompassing hundreds of vehicles and people. If there was something more interesting elsewhere in the footage, I missed it. There is simply no way a single human brain, or even a large team of human brains, can process all of the information from an all-seeing eye.

That being the case, there is today an ongoing quest to build artificial intelligence that can do the watching for us. Such a machine is as close as anything to a holy grail in the world of spycraft. If a camera that watches a whole city is smart enough to track and understand every target simultaneously, it really can be said to be all-seeing; it may even be capable of predicting events before they happen. And when such a camera and its computers find their way into new domains, and become fused with the surveillance systems in all other areas of modern life, what you get is nothing short of omniscience.

For those behind the cameras, the totalist approach to surveillance is a boon. For those on the ground, like the occupants of the blue car, it leaves little room to hide. To be sure, aerial surveillance can certainly be used for purposes we can all agree upon, from firefighting operations to disaster relief. But there is a very real line beyond which the all-seeing eye becomes a dragnet that is incompatible with the tenets of civil liberty, particularly if the one doing the watching is a computer. The billion-pixel question is, where do we draw that line?

With this book, I wanted to give both the technology's proponents and detractors an opportunity to air their arguments as to why we should or should not welcome our panoptic future with open arms. These voices include the engineers who developed WAMI, most of whom have never before publicly discussed the implications of their creation. Some seem almost remorseful about what they have brought into existence. Others, less so. But even the technology's most ardent champions agree that it has significant dangers.

If we want to safeguard the precious little privacy we have left against those dangers, we need some ground rules. As you'll see, I have laid out a blueprint for what those principles and rules might be, not only for WAMI, but for all the forms of wide-area surveillance technology discussed in this book: all-seeing ground cameras, fusion systems that cross-check the footage from those cameras with social media posts, cell phone trackers that can pick up an individual's every move even when they're indoors — the list is a long and harrowing one.

There is ample evidence from recent history to suggest that it is possible to rein in any new tool of surveillance before it gets the better of us. But we need to act soon. The technology is real — we're not talking about flying cars or mind control — but it still isn't too late to define, concretely and proactively, how it should and shouldn't be used. Otherwise, if left unimpeded, wide-area motion imagery will only become more widely used and, yes, abused.

The task at hand hinges, first and foremost, on understanding what, exactly, the all-seeing eye *is*. Hence this book. It's time to stare back at what's watching us from above. Starting with the remarkable story of how it got up there in the first place.

PART I

The Origin of the All-Seeing Eye

1

A New Threat

ON MARCH 27, 2002, seven months after the invasion of Afghanistan, US Navy SEAL Matthew Bourgeois was killed by what appeared to be a land mine as he conducted an ordnance-disposal operation at Tarnak Farms, a compound near Kandahar that had until a few months earlier served as the residence of Osama bin Laden.

He was the 31st US service member to die in the war, but the first to fall victim to a new hazard that would come to define, more than any other weapon, American warfare in the early 21st century.

The following year, on March 29, 2003, four US soldiers died when an insurgent detonated 100 pounds of C-4 in the trunk of a white taxicab at a checkpoint near the city of Najaf, about 100 miles south of Baghdad. Two months later, Army Private First Class Jeremiah D. Smith was escorting heavy-equipment transporters along a road in Baghdad when an object exploded under his vehicle, killing him instantly.

Something unexpected was happening. It didn't even have an official name—in the Pentagon report, Smith's death was attributed to "unexploded ordnance"—but it soon would. No other adversary weapon in the wars of the past two decades has done more to shape US military tactics, technologies, and policies than the improvised explosive device, better known by the acronym "IED."

Very quickly, it became clear that the US military was ill equipped to protect its soldiers from the bombs themselves, and unprepared to fight the elaborate insurgent networks behind the attacks. Seven months into the war, General John Abizaid, head of the US Army Central Command, wrote a classified memo to Defense Secretary Donald Rumsfeld and the chairman of the Joint Chiefs of Staff, warning of the potentially catastrophic effect of widespread IED use. If the United States and its coalition partners were going to lose the war in Iraq, Abizaid predicted, the IED would be the reason.

To stanch the bleeding, it was clear that — as with every major threat the US military had faced since the First World War — airborne surveillance would need to be a part of the response.

Just Watch

Up until the US invasions of Iraq and Afghanistan, most of the imagery that the Pentagon and the intelligence agencies collected from the sky consisted of still photographs — a medium that would have been familiar to any pilot who had flown over the Western Front in 1915. Even pilots who flew the F-117 Nighthawk, the military's most advanced fighter jet, identified their targets using a binder of black-and-white printed satellite images, according to Steve Edgar, a retired Air Force pilot who flew the stealth fighter in the late 1980s and early 1990s.

These still photos, produced by an arsenal of spy planes and satellites that would periodically revisit each target on America's lengthy list of areas of interest, were sufficient for the Pentagon's Cold War needs. The type of targets the United States was most concerned about — nuclear submarine bases, missile silos, and so on — tended not to move. If you took a photograph of an adversary airfield on a Tuesday, it would still be there during a bombing raid on Friday. Even those targets that did move, like a column of tanks plodding along a highway, would generally pursue a predictable course at a predictable speed.

This is how, by using little more than black-and-white photos, many of which were several months old, Nighthawk pilots achieved a 75 percent hit rate during the Gulf War, a feat described by an official record as "unparalleled in the history of air warfare."

By the early 2000s America's primary enemy was no longer a country, but a disparate group of stateless combatants who moved constantly and unpredictably. If a satellite took a photo of an al-Qaeda leader at breakfast, by the time it revisited the area at lunchtime he was long gone. Using traditional airborne surveillance against these groups would be like trying to follow O.J. Simpson's white Ford Bronco by taking a photo of the entire city of Los Angeles every hour and a half. As James Poss, a US Air Force major general who spent close to three decades working in intelligence operations, liked to say of traditional airborne surveillance: all anyone needed to escape the US Air Force was an ignition key — that is, access to a motorized vehicle.

In April 2001, sensing this changing tide, Donald Rumsfeld called for the Pentagon to evolve from its outdated Cold War footing. Instead of keeping America's small number of large, expensive intelligence-collection systems pointed at a few known enemies, Rumsfeld wanted to have many sensors looking at everything all the time, both in the battlefield and in civilian areas: an eye that never blinked.

To do this, his office noted in a report, the Pentagon would need to accelerate the development and fielding of advanced unmanned aircraft. One such aircraft was the Predator, the drone that would later become the emblem of remote warfare.

At the time, the Predator was neither well known nor revered. A fragile motorized glider made by a California company better known for building nuclear reactors, the Predator had, until three years earlier, been considered an experimental aircraft — and an unpopular one, at that.

It flew at only 70 miles per hour, and in its early deployments in the Balkans in the mid-1990s, Serbian ground forces had shown that it was easy to shoot down. Pilots were understandably chary of swapping the cockpit of an actual airplane for a windowless ground-control station equipped with joysticks and computer monitors. Many of those who volunteered to "live life at 1 G" often did so grudgingly after missing out on coveted assignments in manned aircraft units. Worse still, the Predator was unarmed. Its primary payload was a video camera, mounted in a round gray pod bolted under its nose.

But those who had spent any amount of time around the Predator knew that its camera — technically two cameras, one for daytime and one for night — was indeed something of a weapon, and a deadly one, at that. Since it captured moving footage, targets with an ignition key could no longer escape detection. And because the Predator did not need to land to replace its pilots, it could remain airborne for 24 hours — it never blinked. Chances were that a Predator's quarry would have to stop somewhere to rest before it did. (The original un-

An MQ-1 Predator drone on the ramp at Southern California Logistics Airport during a training operation in 2009. *The appearance of US Department of Defense (DOD) visual information does not imply or constitute DOD endorsement. US Department of Defense/Paul Duquette*

armed Predator was subsequently equipped with Hellfire missiles in 2001, but to this day many consider its surveillance powers, not its lethal ordnance, to be its most transformative feature.)

What Rumsfeld didn't share when he called for a new paradigm of post–Cold War surveillance was that the government was already more than a year into its first experiment in bringing the Predator to bear on this new generation of enemies. In a classified operation in 2000, the CIA had flown a Predator over Afghanistan to track Osama bin Laden, the original ignition-key target.

It would later emerge that the Predator had come within a hair of turning the course of history during that operation. At one point, the Predator had found bin Laden strolling across Tarnak Farms in the open air. Unlike surveillance satellites or spy planes, which were limited to photographs, the Predator was able to take up an orbit over the area and wait for the Navy, guided by the Predator's video feed, to dispatch a cruise missile to the compound in order to take out the al-Qaeda leader and his closest associates. But for reasons that would likely seem baffling with the benefit of hindsight, a decision was made to hold off on pulling the trigger. It was America's last open shot at al-Qaeda's leader before 9/11—an opportunity made possible by a single video camera on a drone that nobody wanted to fly.

"What Did We Miss?"

As soon as the IED became a known quantity in late 2003, the Air Force was eager to put its small number of Predators up against the problem. With its high-power video cameras that could reportedly make out an earring glimmering in the sun from 20,000 feet, the Predator had already proven in unclassified operations in Afghanistan that it was particularly useful in helping troops on the ground steer clear of danger.

But the chances of a Predator actually finding a buried IED or a group of "emplacers," a term that refers to those individuals responsible for installing the device, were slim at best. Even the most ambitious efforts to beat the IED in this way fell short. In an operation in 2004 called "the Blitz," the Pentagon dedicated several Cold War spy planes, dozens of drones, and a number of tethered blimps to a single 20-kilometer stretch of road in Baghdad. But it made no discernible difference to the rate of IED blasts and resulting casualties.

Every so often, a Predator crew tracing their camera along a roadway would get lucky and spot an incongruous pile of garbage—a common sign of an IED—or, even more rarely, a group of emplacers digging a hole. But these were flukes.

For those Predators assigned to watch over ground troops on patrol, it was, however, common to bear witness, in exquisite digital detail, to the devastation that a roadside bomb could wreak on a convoy of soldiers or a crowded civilian thoroughfare. Watching these scenes, a pilot's instinct might be to look away. But "your job," Brad Ward, a former Predator pilot, told me, "is to keep staring and to keep watching."

Because of their unique perspective, Predator crews felt a heavy burden of responsibility when those on the ground were injured or killed. After witnessing an attack, drone crews would pore over video from the minutes leading up to the attack. "What did we miss?" they would ask. It was always crushing to discover, all too late, the telltale patch of distressed earth where the IED had been buried. "Oh," they'd say. "We missed *that*."

The Pentagon and the intelligence agencies soon realized that instead of trying to find every IED planted in the ground before it went off, they should seek, and then kill or apprehend, every member of the team who put it there.

To better watch its new mobile targets, the Air Force had developed a rotation system that allowed its Predators to focus on or "stare" at a single target indefinitely without blinking — if need be, for days or even weeks on end. As one drone began to run out of fuel, a second drone would come in to replace it. The pilots on the ground would rotate in and out — unlike with manned aircraft, this was simply a matter of switching seats — as would the intelligence analysts watching the feed.

With this continuous view, a new doctrine of airborne intelligence emerged, called persistent surveillance. The premise was that you can rotate as many drones and crews as you want, James Poss explained, "but the eyes don't go off the target."

These surveillance missions would run as long as necessary. The US Air Force spent 630 hours watching Abu Musab al-Zarqawi, the elusive leader of al-Qaeda in Iraq, before killing him with a strike in 2006.

If you watched a target enough, you learned everything about him — what time he gets up, where he goes to take his coffee in the morning, what time his guards rotate out, whether he has a family. "Is there one guy that gets in the white pickup truck all the time, or is it two guys all the time?" said General Michael Wooley, who headed the Air Force Special Operations Command from 2004 to 2007, in a rare interview near the end of his tenure. "Is there a dog in the back of the pickup truck all the time or occasionally? Do [the men] both smoke? Do they smoke with their left hand or their right hand? Do they hang their elbows out the window, or do they have the window up? Do they flick their cigarette out the window, or do they put it out in the ashtray?"

Ward, who flew some of the very first persistent-surveillance missions, called it "the God's eye view."

Crews would get to know their targets so well, Ward said, that it was not uncommon for the operators to develop an odd kind of affection for the pixelated characters on the screen. Some of the younger operators would give the targets nicknames. The target who smoked a pack of cigarettes a day might be Cancer Man; the one who had to stop and pee every thirty minutes, Tiny Tanker. Ward didn't like this. "Don't give them nicknames," he remembers telling some of the younger pilots. "Because you may watch them die."

The Soda Straw

In the early years of the wars, these drones, along with dozens of teams of special-operations units on the ground, had focused their efforts on a small number of very senior leaders. They were like "spear fishermen," explained General Michael Flynn in a 2015 interview. (A decade before becoming embroiled in Special Counsel Robert Mueller's investigation into Russian interference in the 2016 presidential election, President Donald Trump's former national security adviser had served as the head of intelligence for the Pentagon's special-operations forces.) Every senior leader killed in a strike or captured in a raid in Iraq and Afghanistan was summarily replaced, and the intelligence community began feeling like it was engaged in a morbid game of whack-a-mole. "What we really needed," Flynn said, was "net fishermen."

For one simple technical reason, Predator drones were not good net fishermen. To understand why, imagine you are watching a football game from the nosebleed section of a very large stadium. You are the Predator's operator; the 100 yards of turf is the battlefield. Like a drone, you have a single video camera to record the action. This presents you with a dilemma. Should you zoom in closely and follow the ball as it moves from player to player, or keep the camera zoomed out, with a view of the entire field?

With the camera zoomed in, you can certainly capture the ball's journey in great detail, but you will miss everything else that happens on the field. If, on the other hand, you zoom all the way out in order to watch all the action at once, the players are so small on the screen

Scott Swanson, the chief pilot for the CIA's surveillance operation against Osama bin Laden in 2000 and the first US airman to fire a missile from a modern drone in active combat, poses with the Predator's camera turret, arguably the drone's most transformative weapon, at Tuzla Air Base, Hungary, in the fall of 1998. *Courtesy of Scott Swanson*

that you can't even distinguish your team from its opponents. This is referred to as the soda-straw problem, because watching a target with the camera zoomed all the way in gives operators such a narrow field of view that it can feel like looking through a plastic straw.

In war, the soda-straw problem can be a serious, sometimes lethal handicap. A leaked 2011 Scientific Advisory Board study concluded that the narrow-field view was one of the primary drawbacks of advanced drones. Say a Predator crew is watching a group of people walking down the street. If you want to figure out whether they are military-aged males, you have to zoom in. But if the group is large, you won't be able to see all of them at once. So you zoom out again, but then you may not be able to tell whether the object one of the men is carrying is a gun or a garden hoe.

Operators can switch between zooming in and zooming out, but each option carries the risk that the operators may miss something important. Scott Swanson, a Predator pilot who was the chief pilot for the CIA's aerial-surveillance mission tracking bin Laden in Afghanistan and went on to become the first operator to fire a missile from a Predator drone in combat, describes the challenge of knowing when to zoom in and when to zoom out as "part art, part luck."

In one operation that was described to me, intelligence analysts using a Predator to track a convoy carrying a significant target of interest faced a nerve-racking decision when the vehicles split at a junction and began heading in separate directions. Not knowing who was in which vehicle, the operators had to pick one of the cars on little more than intuition. In the end, they turned out to be wrong and the target was lost. This was not an uncommon situation.

Even following a single vehicle could be difficult with soda-straw vision. When a vehicle approached a T intersection surrounded by tall buildings, the operators would have to anticipate which way it was going to turn in order to position the drone such that it didn't lose visual contact. Anticipating the possibility that they might lose sight of a target behind a building, crews would take note of the pattern of hot spots on the vehicle's hood — the areas of higher temperatures that could be seen by the drone's infrared cameras — so that it could be picked out from among the other traffic once they regained a view of the road.

Generally, Predators would keep their cameras tightly zoomed in on their targets, but to push the football analogy a little further: individual players do not win and lose football games — teams do. With the camera zoomed in, you miss the larger patterns and strategies of play, the broader scheme of things, the network of players. Even if it kept its camera all the way zoomed out, a single Predator assigned to watch over a 40-square-mile area would be able to spot — much less follow — at most only one in twenty moving vehicles. The Predator crews watching Zarqawi found that he was associated with 17 different locations around Baghdad. With his "net fishermen" analogy, Flynn was suggesting that such associated locations should also be targets. But if the Pentagon had decided to watch all of those places persistently, even if they were just a few blocks apart, it would have needed 17 individual rotations of Predators, more than the Air Force's entire inventory at the time.

A New Manhattan Project

In the fall of 2004, while Air Force colonel Steve Suddarth was assigned to Strategic Command, the organization responsible for the US nuclear missile program, he was ushered into a tiny office hidden deep inside the Pentagon. An engineer by training, Suddarth is tall and pale, with a soft, kind face and a disarmingly gregarious manner that would easily belie the gravity of his job.

Suddarth's handler for the day, an engineer named Hriar Cabayan, closed the door to the office and produced a classified document. It was the 2003 memo from General Abizaid to the secretary of defense and the Joint Chiefs of Staff, outlining the danger that IEDs posed to coalition efforts in Iraq and Afghanistan. Cabayan explained that Abizaid had called for a "Manhattan-like project" to deal with the IEDs. In Abizaid's estimation, it was a problem that matched the "complexity and urgency," as one Army report put it, of the nuclear threat. In response to the memo, the Pentagon had created the Joint Improvised Explosive Device Task Force, a secretive enterprise that would work exclusively on the issue.

Casualties from IEDs were continuing to mount and the senior command was willing to try anything that had even a remote chance of saving a life. Abizaid had told his staff a few months prior to Suddarth's visit that any technology with a "51 percent" chance of success qualified for deployment to the battlefield. Soldiers in Iraq had taken to tying leaf blowers to the bumpers of their Humvees to clear the roadside trash that was often used to conceal planted bombs.

At the time, Suddarth was serving as Strategic Command's chief liaison officer with the Department of Energy's national laboratories, the research complexes that work primarily on nuclear-weapons-related projects. According to Suddarth, Cabayan believed that if any organization could make a Manhattan-like contribution to the war effort, it would be the national labs that had emerged from the Manhattan Project itself.

Suddarth left the Pentagon determined to do his part. He began by surveying the developmental technologies that were being put forward by government research laboratories and defense contractors around the country. The "51 percent" benchmark had drawn a bevy of little-known defense firms looking to capitalize on the Pentagon's eagerness to try untested technologies. One of the more infamous products to emerge from the project was the Joint IED Neutralizer, a much-hyped short-pulse laser cannon that could allegedly defuse the devices at a distance. Some task force officers reportedly described it as "bullshit," though not before the contractor, a little-known Arizona outfit called Ionatron, was awarded millions of dollars in contracts.

The task force — which would later become the Joint Improvised Explosive Device Defeat Organization (JIEDDO) — would go on to spend over $75 billion on counter-IED technologies, many of which were futuristic and untested: laser-induced breakdown spectroscopy systems, which use plasma to detect the ammonium nitrate in explosives; specialized drones; ground-penetrating radars; electromagnetic detection systems; and so on.

Suddarth relishes in the details of his work. He is often given to long, meandering technical ruminations in response to simple questions. He approached his new assignment with an intense fastidiousness. In total, he examined over a hundred different technologies. Even when I spoke to him in 2016, a dozen years later, he refused to discuss any of them in detail — except for one. Suddarth had heard that a group of engineers at Lawrence Livermore National Laboratory were working on a surveillance system they claimed would be ideal for rolling up insurgent networks. The moment he saw it in person, he knew he had found what he was looking for.

2

Enemy of the State

ONE FRIDAY EVENING in the winter of 1998, a researcher at the Lawrence Livermore National Laboratory, the nuclear-research facility about an hour's drive from downtown San Francisco, went to the local theater with his wife to see *Enemy of the State*. The choice of movie might have been somewhat odd for a date night, but it was a decision that proved to have historic consequences.

In the film, a rogue cell within the National Security Agency embarks on an illegal operation in pursuit of a labor lawyer named Robert Dean, who has unwittingly taken possession of evidence linking the spy agency to the killing of a congressman who had refused to support a bill that would have granted the government wide-ranging powers to spy on American citizens. As Dean scuttles around the Washington, DC, area, not entirely sure why, or by whom, he is being chased, the NSA deploys a dizzying arsenal of surveillance technologies. Agents tap into his calls at the click of a button, bug his wristwatch with a tracker, and hide a camera in the smoke detector in his living room.

The government's ultimate weapon is a giant video-surveillance satellite that it stations over the eastern seaboard for the duration of the action. Zooming in on the DC area, the satellite beams down a crisp, stable view of the ant-like Dean and his associates. It tracks Dean's nanny as she drives to pick up his son from school. It records his meeting with a mysterious stranger on a rooftop, and it follows him in a blue El Camino as he speeds across the state of Maryland. As far as the movie's plot is concerned, the satellite can see everything. In an interview, the movie's director, Tony Scott, called it the NSA's "Big Daddy."

Such satellites didn't actually exist at the time, and the movie makes a compelling case for why we might be wise to keep it that way. But watching the film with his wife, the Livermore researcher found something else: inspiration. (For security reasons, he requested that his name be withheld.) Whereas his fellow moviegoers saw a herald of doom, he saw an opportunity. *What if such a device could actually be created?* he thought. *Wouldn't that be amazing?*

Lawrence Livermore was founded in 1952, at the behest of Edward Teller, the Hungarian nuclear physicist who was instrumental in the design of the first hydrogen bomb. Initially, it was tasked with the development of thermonuclear warheads that could help the United States keep pace with Russia's rapidly expanding strategic arsenal. The lab had been responsible for the Minuteman missile and the core elements of the Strategic Defense Initiative, better known as the Star Wars program, which sought to develop satellites that could shoot down nuclear warheads before they entered US airspace.

By 1998 the lab had expanded its mandate beyond the nuclear arsenal, but it remained focused on the country's grandest and most sensitive national security and intelligence programs. Taking a fictional Big Daddy satellite and turning it into a (secret) real-life weapon would come right under its bailiwick. After the movie ended, the researcher rushed home and called his program manager. It was late, so his call went to voicemail. "I have a great idea," he said breathlessly. "Call me."

On Monday morning, a number of the researcher's colleagues convened in an office at Livermore, and the researcher explained his idea in detail. *Imagine all the things that the government could do with a wide-area persistent video-surveillance satellite.* But many in the room were unmoved. Sure, they said. It would be great, if it wasn't totally impossible — which, they assured him, it was. A satellite watching such an enormous area would itself have to be immense, and it would generate far too much data for any computer to process.

One colleague, however, offered a dissenting view. Perhaps it wasn't possible with currently available technology, but given the ongoing advances in digital cameras and computer processing, it might be possible in the not-too-distant future. Perhaps *it was worth looking into the matter a little further.*

With that, a small subset of the group launched an in-house program to explore — theoretically, at first — how emerging digital-imaging technology could be affixed to a satellite in order to produce something like *Enemy of the State*'s Big Daddy.

Film the Whole Thing All the Time

As it happened, the group wasn't alone among the Livermore community in its dreams of building a fanciful surveillance satellite. In 2001 they were approached by John Marion, a lanky, soft-spoken engineer with a round, youngish face who was working with Edward Teller, still an active presence at the lab at the age of 91.

A decade earlier, as part of the Star Wars program, Teller had proposed a project called Brilliant Eyes, which involved launching a constellation of several hundred small, cheap satellites that would orbit the Earth at an altitude of 300 kilometers, arranged in such a way that "all the earth could be watched all the time." At a conference in 1992, Teller suggested that the information collected by these satellites could be put toward "peaceful applications in everybody's interest," such as providing communications coverage in remote areas, warning of natural disasters, managing air traffic, and reporting on the weather.

Though the US government had largely abandoned Star Wars and with it Brilliant Eyes by the time *Enemy of the State* was released, Teller had not. Working with Marion, Teller was now developing a blueprint for a global weather-tracking system that involved unleashing many thousands of small balloons into the atmosphere and then watching them with dozens of satellites.

In the spring of 2001 Marion had attended a briefing presented by the digital-photography group, which included a clip from *Enemy of the State*. He realized that their project was based on the same principal as Teller's, and he joined the project shortly thereafter.

The logic of Cold War aerial photography was to capture the widest possible area. The Predator's mandate was to stare, narrowly and unblinkingly, at a small number of individual enemies. The Livermore team's idea was, Why not do both? Build something capable of achieving the same wide-area coverage as a traditional still-photography satellite, but that recorded video with enough magnification power to watch every individual moving target within that area simultaneously.

Being nuclear laboratory researchers, when the team members considered possible applications for such a technology, their minds first turned to nonproliferation operations. At the time, Saddam Hussein was mostly refusing to cooperate with weapons inspectors looking for evidence of weapons of mass destruction (WMD) at several Iraqi facilities. Meanwhile, colleagues from another department at Livermore were preparing a classified report showing that North Korea was launching a new uranium-enrichment operation with the assistance of Pakistani engineers. For its part, the CIA had begun to collect reports that North Korea was, in turn, coordinating with Syrian engineers and officials, who would later go on to build a nuclear reactor at Deir ez-Zor.

The facilities where these alleged activities were thought to be taking place were known, but the activities themselves were difficult, if not impossible, to confirm. What could you find out, Marion and his colleagues wondered, if you put their proposed camera right over one such "enigma" facility and just watched? "If you have an area where you know bad stuff is going to happen," Marion told me, describing the group's thinking, "but you don't know what's going to happen, or when, or who, or what they're going to do, the only solution is just to film the whole thing all the time."

One could follow the facility's staff members leaving the site and track them to their homes, revealing their identities and, by extension, their educational and professional backgrounds. If they had degrees in nuclear physics, there was probably something suspicious about their employment at a purported lightbulb factory. Because the satellite would record the whole area persistently, one could also play the recorded footage backwards. If a truck entering the facility were tracked back to a uranium mine, one could make a reasonable guess that the facility was not, as claimed, a fertilizer plant.

Marion pitched the idea to a former colleague at the National Reconnaissance Office, the agency that managed the US spy-satellite constellation. But the budget was steep—$50 million just to build a functioning prototype. According to Marion, the NRO official worried that if he backed Livermore's effort, he'd lose his job.

One of Lawrence Livermore's earliest wide-area camera prototypes bolted to the nose of a civilian Eurostar AS350 helicopter. This aircraft flew secret test missions over various locations in California, including downtown San Diego. *Logos Technologies/John Marion*

Undeterred, the team pitched the Department of Energy, which agreed to provide a smaller budget for further research. The funding would not be sufficient to build a whole satellite, but it was enough to build a camera.

From the outset it was clear that such a camera would have to be larger than anything in existence: it would have to be capable of taking in an entire city-sized area in one frame with enough resolution to detect individual cars coming to and from the center of the action. The team had originally flown a professional survey camera out of the side of a police helicopter over Santa Rosa. Though that system was capable of capturing a wide area, it could record only one frame every ten seconds, too slow to track an ignition-key target.

As a second attempt, with the funding from DOE, the researchers wired together two 11-megapixel video cameras. It wasn't a polished unit, but it could take more frames per second, and it was at least ten times more powerful than a Predator's cameras. It was also relatively inexpensive. A German company called Vexcel later built a single camera of roughly equal capabilities, but it cost $800,000 — by comparison, Livermore's home-brew unit cost just $80,000.

To replicate the steady, top-down view as seen from the satellite in *Enemy of the State,* the Livermore researchers decided to go to the source. They reached out to Wescam, the company that had run the aerial filming for the movie. The crew, it turned out, had created the "satellite" imagery by flying a helicopter at high altitude with a film camera pointed directly down at the earth.

Nathan Crawford, a general manager at the company, invited the engineers to meet in Los Angeles, where he was working on the set of *Terminator 3.* The engineers watched Crawford's team film a complex stunt scene using a camera mounted on a tall crane. They were deeply impressed. Crawford had already been involved in a number of classified technology projects at Wescam's defense unit, which had developed the Predator's stabilized camera-mount. Thanks to a lengthy career in camera work, he also possessed the rare ability to film a golf ball flying through the air at great speeds — a useful skill if you wanted to emulate a satellite tracking a cohort of nuclear scientists going about their daily routines from several hundred miles above the surface of the earth.

The new, improved camera first flew in a series of tests over the San Fernando Valley in the fall of 2002. In one flight simulating the arc of a satellite in orbit, Crawford, sitting in a helicopter, climbed from 1,000 feet to about 15,000 feet while keeping the camera fixed on a Shell gas station. The task was much like filming a golf ball on a long tee shot, except that in this case he was the one soaring through the air, and his target was fixed in place on the ground.

Attack the Network

When the United States confronts a challenge of Manhattan Project–like proportions, it is likely that a little-known group referred to as the Jason* Defense Advisory Panel will have a hand in shaping the response, often before the general population understands the nature of the problem, or even that the problem exists. Formed in 1960, Jason consists of a rolling membership of eminent civilian physicists,

---

*Named after the Greek hero Jason of the Argonauts.

chemists, biochemists, mathematicians, and economists — including at least eleven Nobel laureates over the years, at the most recent count — who periodically meet to evaluate a highly complex topic of great national security import and develop recommendations for how the defense and intelligence establishments ought to go about addressing it.

Though Jason rarely makes headlines, the group has had a remarkable influence on many headline-grabbing topics. One early Jason study concluded that the use of small nuclear weapons against North Vietnamese infrastructure "would be uniformly bad and could be catastrophic." Another endorsed a startlingly expensive — and largely unsuccessful — effort that came to be known as Igloo White, which sought to track North Vietnamese forces moving along the Ho Chi Minh Trail using advanced remote-sensing techniques. More recently, the group has weighed in on the merits of plasma-gun technology, simulated the impact of severe solar storms on the US electric grid, and provided guidance for the Human Genome Project. Its goal is to stay ahead of the curve; in 1991 it met to assess the potential of remotely piloted aircraft and small satellites, two technologies that wouldn't become mainstream for another two decades.

In the fall of 2003 the CIA and the Pentagon's research and development directorate turned to the Jason Defense Advisory Panel for help on the IED problem. The Jasons agreed to conduct a joint study on the topic, with an eye to endorsing any technologies and strategies it identified that appeared to be particularly promising. The CIA's chief scientist, John Phillips, assigned the task of coordinating with Jason to the agency's Dan Cress, an expert in seismic, acoustic, and electromagnetic surveillance who was concurrently working on an assignment addressing Syria's WMD program.

In the Jasons' first meeting on IEDs, held in La Jolla, California, in January 2004, a series of briefers described how improvised explosive devices work, identified some of the groups that appeared to be responsible for planting them, and explained why they presented such a severe threat to the US strategy in Iraq and Afghanistan.

The group also saw presentations on a variety of tools designed to detect and neutralize improvised bombs. Cress was unmoved. As an intelligence officer, he didn't see it as the CIA's job to select the best steel plating for a Humvee's underside. The intelligence community's goal was to find the terrorist cells *behind* the IEDs. He decided that the second meeting, to be held three months later, should focus on how to dismantle these groups, rather than going after the explosives themselves — a strategy known as "attack the network."

When Cress spoke to John Phillips about whom to invite to the second meeting, the CIA's top scientist described a surveillance project that he had learned about during a recent visit to Livermore. Phillips said that he had sat in on a presentation by an engineer named John Marion, who described an interesting idea for a surveillance satellite.

Marion still had not found a sponsor organization to provide the funding necessary to build a device that might actually be used in real operations, though not for lack of trying. In a single 12-month span, Marion — who was normally based in California — clocked over 100 nights in the Key Bridge Marriott, a ten-minute drive from the Pentagon. Casting about for new potential applications for the technology that could interest sponsors beyond the nonproliferation community, the team struck on the idea that a wide-area camera might be just as useful against a terrorist network as it would be against a team of foreign nuclear scientists.

A briefing by Marion was a regular fixture on the Lab's tour agenda for visiting officials from other agencies — it was seen as a way of introducing the officials to some of the cutting-edge work happening in his department. In late 2003, shortly before Phillips's visit, Marion had added a few slides to his presentation describing how his team's camera could be used against a network of insurgents.

When Cress saw the presentation for himself, on a visit to Livermore the following March, he invited Marion to attend the second Jason meeting, right on the spot.

At the meeting, which was held the following month, Marion sat in silence as the Jasons worked over the attack-the-network problem. Many of the Jasons appeared to be unsatisfied with the options that were brought before them. There were multiple disparate IED cells in each of Iraq's major cities, but they were exceedingly difficult to find and track.

At one point in the proceedings, one of the Jasons grew exasper-

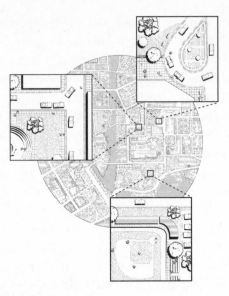

Wide-area motion imagery records a very large area in such high resolution that users can zoom in on areas of interest while the camera continues recording the entire view. *Emily Wissemann*

ated. If only there were a way to surveil the adversary's *entire area of operations* all at once, he said, it would be much easier to connect the perpetrators to the attacks. Marion was the next speaker called up to present.

IEDs were going off in Baghdad almost every day, Marion explained; if you filmed a large portion of the city with a wide-area camera, you were bound to catch an explosion in your frame. You could then download the imagery and watch the incident. From the point of the blast, you could find the people who had installed and triggered the IED and track them backwards to where they had come from simply by rewinding the tape.

Predator crews already employed this strategy whenever they were lucky enough to chance upon an IED emplacement team in the process of planting a bomb. Instead of killing the group on the spot with a Hellfire missile, they would use the Predator's other weapon — its camera — to follow the figures on the screen in the hope that they would be able to lead them back to other, more senior members of the network. With a wide-area camera, one didn't have to rely on a chance encounter with an IED team to begin such an investigation.

Once you had backtracked a vehicle to an address, you could watch that house as though it were a suspected nuclear facility, tracking all the vehicles and people that came and went. Hidden in plain sight among the general population, each IED cell consisted of up to eight people, including triggermen, bomb makers, financiers, emplacers, and a cameraman. Each vehicle (a "proxy" in spy parlance) that stopped at the location (sometimes referred to as a "node"*) could, in turn, be tracked both forward to its next stop and backward to the place it had come from, thus revealing new nodes in the network. With a wide enough camera, there was no soda-straw problem. If two cars split up at an intersection, you could follow them both, as well as hundreds of others, with a single airplane.

To prove they could follow vehicles in this way, the Livermore team had flown the camera in a helicopter over an oil-storage facility in the desert south of the town of Mojave. Actors in two vehicles had driven in a series of patterns between two separate areas of the facility. During the test, Marion told the facility's security guards that he was filming a movie, which was technically true. When one of the guards asked where all the cameras were, Marion pointed to a small speck in the sky: "Up there."

Any vehicle that visits a node would become a new proxy, and any location that those new proxies visit would become a node, and so on. As one former senior intelligence official, Keith Masback, told me, if you're dealing with a true network, patterns will soon emerge. Proxies involved in separate incidents might visit the same location. One proxy might travel between two known significant nodes, connecting a local insurgent cell to a broader regional organization.

Cress and the Jasons were rapt. Here, finally, was someone speaking the intelligence community's language. With a full picture of the group's structure—an "association matrix"—the intelligence agencies needn't rely on the whack-a-mole strategy. Instead, they could

---

* A term that can also, somewhat confusingly, refer to individual members of a network.

use subtler, more devastating techniques. Using what's known as "social network analysis" to map out relationships among members of a group, one could, according to a detailed Pentagon instruction manual known as the *Commander's Handbook for Attack the Network*, identify particularly crucial members of the group who "can be killed, captured, or influenced to achieve the desired effect" of disrupting its operations.

One Jason member, the eminent physics professor Roy Schwitters, told me that he will never forget Marion's briefing. The group decided to immediately draft a classified two-page letter endorsing the wide-area-surveillance concept, and sent it to John Phillips of the CIA and Ben Riley, who worked for the Pentagon's Rapid Reaction Technology Office, a program established shortly after 9/11 that focused on building and deploying weapons for counterterrorism operations as quickly as possible.

Shortly after the Jasons' endorsement was issued, Cress arranged a meeting with Phillips and Riley. Cress had calculated that it would take about $6.5 million to further develop Livermore's technology to a point where it could be offered to one of the military branches for deployment. Phillips and Riley committed to splitting the investment between the agency and the Pentagon. For his unstinting early faith and support, Marion still calls Cress "the Godfather."

The program had caught its first big break, just as the tests themselves began to yield revolutionary results. That same month, Nathan Crawford flew a second prototype camera to San Diego and ran a series of circular orbits over the airport, the harbor, the downtown, and the Marine Corps Air Station in Miramar.

Though the airplane was flying in circles, thanks to a trick of image processing that would prove to be just as elemental to the development of the all-seeing eye as the camera itself, the footage was clear, still, and maplike. Working with a team from the MIT Lincoln Laboratory that had independently taken an interest in wide-area surveillance for IED networks, the engineers had built computer software that rotated each frame "north up" on the screen, and pinned each pixel to an actual point on the surface of the earth, creating a smooth, stable view of the ground.

The resulting imagery had the appearance of a moving satellite image—like Google Earth, but alive. Everything that was stationary on the ground was stationary in the image (unlike regular aerial footage, in which everything moves relative to the video frame), while anything that was actually moving—cars and people—would "pop," as Crawford put it, against the still background. Though vehicles and people only appeared as small specks of a few pixels, that was all one needed to follow them anywhere they went.

Instead of playing a smooth, continuous video like a Predator drone camera, it captured just two frames per second, creating an effect not unlike those old CCTV video clips in which figures appear to teleport by a couple of feet each second. But the coverage was vast, and eerily similar to the scenes from *Enemy of the State*.

Angel Fire

This was the footage that Steve Suddarth saw when he visited Livermore as part of his counter-IED research mission in the fall of 2004. The entire city of San Diego was laid bare—a perfectly still townscape populated by moving cars and people who stood out clearly against the background. Boats could be seen entering and leaving the port. At one point, an airplane flying at a lower altitude skipped across the screen. Suddarth says it was the most amazing thing he had ever seen.

Suddarth was so impressed, he decided his own contribution to the IED effort would be by way of wide-area surveillance. On a church charity trip to Mexico a few weeks later, Suddarth met a fellow congregant who worked at the Los Alamos National Lab, a nuclear-research facility like Livermore, located near Santa Fe. Riding in Suddarth's airplane back to New Mexico, where they lived (Suddarth was a licensed pilot and often traveled in his single-prop Cessna), the two men decided to propose a collaboration between the Air Force and Los Alamos to try to tackle the IED problem with an aerial-surveillance system of their own.

Suddarth felt that Livermore's prototype had a crucial drawback: the imagery couldn't be transmitted live from the sky to the ground. Instead, Livermore had proposed that the imagery analysts would download the videos at the end of each mission, working through the

A screenshot from Lawrence Livermore's test surveillance flight over San Diego in 2004. The USS *Midway*, a decommissioned aircraft carrier that now houses a military museum, is visible in the top left corner. Steve Suddarth, who was shown the footage during a classified effort to find technologies to use against IEDs in Iraq, called it "the most amazing thing I've ever seen." *Logos Technologies/ John Marion*

footage the way police might pore over CCTV tapes following a robbery. As Suddarth saw it, a live transmission system would be much more useful. Not only could you hunt insurgents in real time, but you could also steer your own forces away from danger as they worked through the battlefield.

When Dan Cress heard about Suddarth's idea, he was sufficiently convinced to arrange for $200,000 in seed money from the CIA's Intelligence Technology Innovation Center to begin working on software for the live link. Cress hoped that Los Alamos and Lawrence Livermore would cooperate and collaborate on a single project, with Livermore generating the imagery and Los Alamos building the software.

But Livermore and Los Alamos were famous rivals in the small world of nuclear weapons development — Suddarth jokes that this was the true Cold War nuclear rivalry—and some of those involved believed that Livermore was indisposed to share what was already shaping up to be a revolutionary project that could put an end to the IED problem and, perhaps, even fundamentally alter the way the United States did aerial surveillance. Los Alamos, in turn, had few incentives to put hard work into a program for which it might not get any credit.

So, to Cress's dismay, instead of waiting for test footage from Livermore, the researchers in New Mexico opted to build their own hardware, a Rube Goldberg–like four-camera setup that they named Angel Fire. On its first test, the engineers flew the system over a mall in Albuquerque. At one point, a delivery van darted along a narrow alleyway at considerable speed. In a meeting in a secure room on Capitol Hill in late 2005, Suddarth showed the footage to a group of congressmen, who recoiled with a mix of awe and horror. "All you have to do is write the ticket," one said. It was true, Suddarth conceded. "I knew what vehicle he was driving, I knew where the vehicle came from, I knew where it went, and I knew what time exactly" it happened, "and I knew what speed he was going," he told me. "The only thing I didn't have was a picture of the driver."

Engineers on the Beach

Livermore, meanwhile, was ramping up its own project. With the $6.5 million in seed money, the team built a new prototype camera made of no fewer than six individual cameras strapped together. It was decided that the camera wouldn't go on a satellite—the tests had demonstrated that it actually worked just as well aboard an airplane flying in circular orbits over the target area. They called the camera Sonoma.

Thanks to the Jasons' weighty endorsement, the idea was now being taken much more seriously at the Pentagon. Less than a year after the Jason meeting, Sonoma was adopted by the Army Research Laboratory. Given its potential to address the IED problem, the Army set the effort on an accelerated track toward deployment, and in February 2005 it took full control of development despite objections from Livermore, which, according to various accounts, wanted to continue much of the work in-house and keep its hold on certain proprietary technologies its engineers had built. The Army hired Logos Technologies, a Virginia-based contractor that hired Marion after his relationship with the Livermore leadership had grown acrimonious, to build the final sensors. The entire project was transferred to a hangar in West Palm Beach, Florida, for the last cycle of development, testing, and preparations for deployment.

There was still some hope that the two national laboratory teams would work together, so Suddarth and several researchers from Los Alamos joined the Sonoma team in Florida. Suddarth brought with him an Air Force colonel named Ross McNutt, an instructor at the Air Force Institute of Technology who was teaching a seminar on rapid technology development. Los Alamos had agreed to let McNutt and the students from his class participate in Angel Fire as a hands-on exercise in the principles that McNutt, who had written his PhD dissertation at MIT on rapid technology development, wanted to impress upon them.

Ultimately, the efforts to merge the two programs would be quashed for good. The Air Force Research Laboratory took control of Angel Fire in much the same way that the Army had taken Sonoma, and many of the original engineers, including Suddarth and McNutt, were removed from the program. But for a few weeks in Florida, the two teams worked side by side.

Even with Army and CIA backing, this wasn't a large, polished program. It felt more like a startup. Between each test flight, engineers would write code to patch the system's many bugs on laptops fed by generators. The project barely scraped by financially. Subcontractors like Kitware, which built the image-correction software for Angel Fire, would receive funding spasmodically, $20,000 at a time.

Nevertheless, Nathan Crawford, who had left Wescam to provide services for Livermore and started his own company, Consolidated Resource Imaging, worked to imbue the effort with a Hollywood ethos and aesthetic. The movie industry had a way of staying on schedule that he felt would keep the project on track. He built large, well-appointed trailers for the engineers and stocked the fridges with their favorite snacks and drinks, as though they were celebrities. As a result, field tests sometimes felt like a large film production rather than a classified intelligence operation.

The work itself, however, was more grueling than glamorous. It was common for engineers on board to develop symptoms of oxygen deprivation from the many hours they spent in the test airplane's unpressurized cabin during tests. Some of the engineers started losing weight. Marion called it "the hypoxia diet."

For the most part, the Livermore and Los Alamos researchers were glad to suffer if it meant getting the job done more quickly. Unlike the many Department of Defense programs that are designed to move methodically and ponderously, like long-distance runners (when you build a nuclear submarine, you want to get every detail right, you want to test and retest, you want to evaluate), this operation was, by necessity, a sprint. On September 14, 2005, a single bombing in Baghdad killed 112 people. Indeed, reports of IED attacks seemed to be the only consistent news coming out of Iraq. These stories weighed heavily on the engineers. Every day until they deployed was another day that attacks would kill service members and civilians. "There started to be a figurative body count on this program," said Crawford.

During that brief period of collaboration in Florida, the engineers tested the camera for hundreds of hours over nearby towns, recording the unsuspecting population below. The workshop was half a mile from the ocean, but most of the team members only made it to the beach once, on their single day off when the test planes were in for repairs.

Sitting on the sand one afternoon, Crawford and Marion talked about what the camera might be capable of beyond rolling up networks of insurgents. They had often discussed how it could be used to patrol ports and borders, or search for survivors of a natural disaster. During the test flights in San Diego, the team had flown the camera over Otay Mesa, a neighborhood that abuts the Mexico border, in a demonstration for US Central Command and Border Patrol. Within hours, the camera had recorded a number of individuals crossing into the United States illegally.

And yet, though Crawford understood the project's potential, he couldn't help but feel a little unsettled. There was something inexorable about the process they had set in motion. The engineers had created a thing of raw, unsubtle power, and it was clear from the outset that it would have consequences beyond the wars in Afghanistan and Iraq. When I first spoke to Crawford on the phone in early 2017, he told me he'd spent the previous fifteen years expecting a phone call from a reporter. "This is a story that has been waiting to be told," he said later. "A lot has happened, and we need to learn from it."

Observe, Detect, Identify, and Neutralize

In early 2006 the cameras were ready. With their 66 million pixels, they were capable of observing an area more than six kilometers wide, thousands of times what a Predator could see through its soda-straw sensor. Mounted aboard a Short Brothers 360 twin-prop cargo plane, one of the few available civilian aircraft large enough to fit the cameras, the operators, and their associated equipment, the system could watch hundreds of vehicles at once.

In order to turn the video into stable satellite-like imagery, the

The original, custom-built six-lens Constant Hawk camera system that was deployed to Iraq in 2006. *Logos Technologies/John Marion*

Lincoln Laboratory had built a 1,000-pound stack of multicore Intel PCs that were "stitched" together with parallel processing software to form a single, giant brain connected to the camera aboard the airplane. This processing unit, according to its creators, was so powerful that it would have made that year's TOP500, an annual ranking of the most powerful supercomputers in the world. But since it was classified, the system wasn't on the list.

The program's secrecy was almost absolute. In a final test flight in February 2006, one day before the system was originally scheduled to be deployed, two aircraft — the program's entire fleet — collided in midair, killing three crewmembers. The local papers and the subsequent accident investigation reported that the planes were flying a seemingly innocuous fuel-tank test at the time of the accident. It wasn't for another 14 years, with the publication of this book, that it would be revealed that the airplanes were in fact preparing to take the first-ever operational wide-area cameras to war.

The program also had a new name. Sonoma, it seems, was too tame for the Army's tastes, so it had been changed to Constant Hawk. (The project had briefly been referred to as Mohawk Stare, because the Army planned to mount the system on the Vietnam-era OV-1 Mohawk, one of the strangest-looking aircraft ever made, but the plan to use the aging Mohawk was eventually scrapped due to a variety of technical and safety concerns.)

Ultimately, five airplanes were dispatched to Baghdad in the summer of 2006, and the program was incorporated into an entirely unpublicized special-operations unit, known as Task Force Dragon Slayer, which was focused on countering IEDs and mortar attacks. The need on the ground was more urgent than it had ever been. US forces in Iraq were encountering more than 80 IED attacks every day.

Like the early Predator, Constant Hawk had its skeptics. Numerous Pentagon officials told Crawford, who deployed with Consolidated Resource Imaging to maintain and operate the aircraft in the field, that it simply wouldn't work. In light of this skepticism, Constant Hawk's initial deployment was only intended to last 90 days, but it performed so well that the aircraft remained active in Iraq until the US withdrawal in 2011.

The Army also began working to build a larger version of the camera along with a second infrared unit for nighttime surveillance operations. The following year, ground forces in Afghanistan requested that Constant Hawk be deployed to their war, too — which ultimately happened in January 2009.

Constant Hawk was capable of exactly the kind of persistent surveillance that had been perfected with the Predator, except that it watched whole neighborhoods at once — not just high-value targets or ground patrols, but the entire environment surrounding them. Whereas a single Predator assigned to a 40-square-mile area could monitor only 5 percent of moving vehicles on the ground, 95 percent of vehicles could be tracked with the wide-area system.

Only a single short clip of Constant Hawk footage was ever made available for me to watch, under tight security at the offices of a defense contractor somewhere in Virginia. But even this one sample made clear the extent to which the life of the city could be captured in its totality under the giant camera's view. In the clip, a long line of cars waits for fuel at a gas station. People can be seen strolling along

A Constant Hawk–equipped Short Bros. 360 airplane shortly prior to deployment at Mitchell Field in Milwaukee in 2006. The multilens camera, at the time one of the most powerful in existence, is visible in the rear cargo door. *Constant Hawk Development Team/John Marion*

the streets. At one point, a herd of goats crosses through the frame. The video is eerily peaceful — one might even call it serene.

That quietness, however, belies a much more violent story. Shortly after it was deployed to Iraq, Constant Hawk was transferred to Task Force ODIN, a new unit created specifically to find, track, and kill insurgents using advanced and experimental airborne technologies. Besides being a reference to Odin, the wizard of Norse mythology, god of wisdom, death, and divination, the name stands for "observe, detect, identify, and neutralize."

"When I do stare, see how the subject quakes," the mad king tells the blind Gloucester in Shakespeare's *King Lear.* Broadly speaking, ODIN's goal was to put fear in the enemy, and it did so with its eyes. Clicking through the imagery frame by frame, analysts at the airfield where Constant Hawk was deployed would follow each suspicious vehicle both forward and backward from the site of an explosion. When the vehicle stopped at an address, they'd pass the location to strike and capture units operating on the ground. Two of the aircraft were equipped with a second, 16-megapixel camera that could take high-resolution close-ups of the targets with the same penetrating acuity as a Predator's soda-straw feed.

Every few days, couriers would personally deliver the data — which was stored on commercial hard drives — to Ramstein Air Base in Germany and an NGA facility in Virginia. The data also found its way to the CIA, where Cress and his colleagues used it to try to develop a fuller picture of the adversary's network (Cress would later say only that the imagery was "helpful"), as well as to a site referred to by one Army document as "Area 59."

Under this new concept of surveillance, the enemy's ignition key went from being an asset to a liability. To illustrate the point, one defense executive discussing operations like Constant Hawk in 2013 quoted the scene in *Jurassic Park* where Dr. Alan Grant faces down a *T. rex*: "Don't move. He can't see us if we don't move." With WAMI, the only way to avoid detection was to keep *very still.*

Though the technology was held together by nothing more than zip ties and "good intentions," as one engineer put it, Constant Hawk soon turned from an experiment into an extensively used prong of the Pentagon's counterterrorism intelligence and surveillance operations. A single variant of the system collected some 10,000 hours of surveillance over Afghanistan in just two years.

As more people tuned in, the intelligence community came around to the idea that a wider view was not only possible, but desirable. After a year, the Pentagon established multiple intelligence cells focused specifically on analyzing Constant Hawk footage in Washington, DC.

Much about Constant Hawk's operations remains classified. Even the airplane's cruising altitude is considered privileged information. One senior commander who worked within Task Force ODIN referred to the aircraft in reverential terms when we spoke, but declined repeatedly to go into specifics. When I asked Crawford to describe the operation, he would only say that the program very quickly succeeded in the role it was intended for: finding the people responsible for planting IEDs and mounting ambushes. "We'd say, 'the guy lives in this house here,'" he told me. "And they'd go knock on his door."

Crawford believes that up to 600 US service members, and countless Iraqi civilian lives, were spared thanks to operations involving Constant Hawk. To be sure, the technology did so by enabling a killing spree of its own. Task Force ODIN is said to have "eliminated" more than 3,000 suspected insurgents, and captured hundreds more, in its first year alone.

3

The Gorgon's Stare

CONSTANT HAWK WAS soon joined in Iraq by Angel Fire, which was deployed in support of a set of Marine Corps units in August 2007. The Marines Corps leadership had requested Angel Fire after watching it in action at a military exercise where it demonstrated the unique ability to track friendly and enemy forces simultaneously. They hoped that it would help stem the heavy losses that they were experiencing as a result of sophisticated ambushes and sniper attacks, particularly in urban zones.

In the field, the system was reportedly beset by technical glitches and many of its features were unusable, but it worked well enough to remain in theater for 18 months. All told, the aircraft flew more than 1,000 sorties over Iraqi cities and beamed some 5,000 hours of live footage down to Marines on the ground. Like Constant Hawk, the Angel Fire program was soon wreaking havoc on Iraqi terror networks. Ross McNutt, the Air Force instructor who worked on Angel Fire, told me that when analysts discovered a location that appeared to be linked to an attack within the airplane's 16-square-kilometer view, they would pass the address to a raid force. Using the same eerie phrase that Nathan Crawford had used to describe Constant Hawk Operations, McNutt said that ground units would move on the address, "knock on the door and say, 'Sir, can we have a moment of your time?'"

As more people heard about Constant Hawk and Angel Fire, many wondered why the two technologies had never been merged. It seemed obvious that they would have complemented each other well. Constant Hawk had more pixels, but Angel Fire transmitted live video straight to the ground, so that intelligence analysts didn't have to wait until the airplane landed to track insurgent networks or snipers waiting on rooftops.

But for reasons entirely nontechnical in nature, it could not be so. Aside from the rivalry between Livermore and Los Alamos, efforts to create a single merged program were marred by internal strife within each team. Marion was fired from Livermore in 2005. Suddarth and McNutt's relationship became strained; Suddarth felt that McNutt — whose technical contributions, Suddarth claims, were minimal — undermined his leadership and took credit for work he had never done, while McNutt, for his part, was frustrated by the plodding, procedural approach of the commanders running what was supposed to be a fast, experimental program. The Air Force Research Lab argued with the Air Force Institute of Technology and Los Alamos. The Army Research Lab clashed with Livermore.

The programs' newfound sponsorship by different branches of the US military only intensified the feuding and further obstructed a merger. The Marine Corps accused various Army officers of trying to obstruct its efforts to acquire Angel Fire as a way of strong-arming the service into investing in Constant Hawk. In response, the Army suggested that the Marines had misled units in the field about Angel Fire's capabilities. In turn, the Marine Corps countered that Constant Hawk didn't meet its specifications. Air Force and Marine Corps officials turned to Congress to keep Angel Fire alive, sending dozens of emails to Senate and House staffers urging them to support their camera over the Army's.

Some in Congress grew impatient. The infighting was slowing down the development process while troops continued to die in Iraq. Running the two projects in parallel was also expensive. The four Angel Fire surveillance systems deployed to Iraq in the summer of 2007 had cost about $25 million. That same year, the Pentagon counter-IED program awarded the Air Force an additional $55 million for Angel Fire and the Army $84 million for Constant Hawk. Senator Kit Bond, a Republican from Missouri, issued a seething public statement accusing the Pentagon's rival factions of "gross mismanagement."

Among those following the saga was Mike Meermans, the staff director of the House Intelligence Committee. A former Air Force Officer, Meermans is a well-known figure in the world of airborne intelligence. In 2014 he was inducted to the shadowy "Intelligence, Surveillance, and Reconnaissance Hall of Honor,"* for his role in a number of classified Cold War aerial-surveillance programs and his subsequent work for the House of Representatives. Among other accomplishments, in 1996 he had convinced his boss, Representative Jerry Lewis, to fund the ongoing development of the Predator, a historic decision that likely spared the program from an untimely termination.

In 2005 Meermans noticed that Constant Hawk and Angel Fire were becoming regular topics of conversation within the intelligence and defense organizations under his purview. Steve Suddarth's Angel Fire briefing on Capitol Hill had left those who attended both alarmed and impressed. Word was that these systems would save hundreds of lives in Iraq once they were deployed. People told Meermans that soon even a *90-megapixel* camera might be possible.

When Meermans and his colleagues tallied the tens of millions of dollars being spent in parallel on Angel Fire and Constant Hawk, they decided to intervene. The group drafted a classified directive for the DOD, according to Meermans: "We're going to stop building two separate systems and we're going to build one."

The new camera, named simply Wide-Area Airborne Surveillance, would combine elements of both systems. It would generate a wider view than either of the existing systems, enough to watch a whole "city-sized" area, and it would send live footage to the ground. It would be optimized for both live overwatch missions and after-the-fact analysis. It would have a secondary infrared wide-area camera. And, unlike Constant Hawk and Angel Fire, it would be mounted on a drone.

After hearing of the directive for what was effectively a wholly unprecedented program — a solution to the Predator's soda-straw problem, no less — a number of senior Air Force officials began lobbying Pentagon leadership for control of the initiative, claiming, among

---

* Established in 1983 by an alumni organization for former Air Force intelligence officers, the Hall of Honor is intended "to pay tribute to individuals who have served with great distinction and contributed immeasurably" to Air Force spycraft. As of 2018, only 192 service members have been inducted.

other things, that it could mount the system on a Reaper, its new strike-capable drone.

One of those officials, Major General James Poss (the same officer who had remarked that in the past all anyone needed to evade the Air Force was an ignition key), had seen a WAMI prototype being demonstrated at a military exercise. He had watched a young lance corporal approach an actor posing as an insurgent. Under questioning, the actor claimed innocence, but analysts rewinding the footage found that he had actually just come from a "radicalized" mosque. Operators told Poss that following a car bomb attack you could backtrack the vehicle to the insurgent safe house where it had been loaded with explosives. *TiVo,* he remembers thinking, drawing a connection to the commercial technology for recording and rewatching TV shows, *combat TiVo!*

After the Pentagon agreed to give the program to the Air Force, an internal planning group, the Joint Requirements Oversight Council, issued Memorandum 106-08, a short document that sketched out performance targets for the new camera. Initially, the exact details of the memo were kept secret even from Congress. Later, a military funding bill called the performance targets "murky." But for the Air Force, it was more than enough to work with. According to people who reviewed the document, it simply called for a camera that could watch more ground — and, thus, more vehicles, more people, and more networks — more persistently, more unblinkingly, than anything in existence.

Big Safari

The situation in Iraq was as bad as it had ever been. Days after the new WAMI program was given to the Air Force, in July 2007, suicide bombers had killed 50 Iraqi civilians celebrating the national soccer team's victory against South Korea in the Asian Cup. The new program was given Quick-Reaction Capability status, meaning that the effort to develop and deploy the technology would be fast-tracked. Under guidelines issued by the JROC, the deadline was set for 18 months.

Many Air Force officials believed the only organization that could build such an unprecedented spy plane in such a short time frame was Big Safari, an Air Force skunkworks that develops modified surveillance aircraft, often for urgent covert and classified missions. Military officials speaking publicly about this highly secretive unit often hesitate to call it by its name, referring to it instead as "a particular Air Force organization."

It was Big Safari that was responsible for modifying and operating the Predator in the CIA's mission against bin Laden over Afghanistan in 2000. The following year, it had armed the Predator with laser-guided missiles with an eye to possibly returning to Afghanistan to kill the al-Qaeda leader.

As part of that effort, a private technical contractor working for Big Safari had single-handedly designed the transatlantic control system that allowed Air Force pilots to fly the Predator remotely from ground stations in the United States. This remote-control system was a temporary fix to a niggling legal issue. Germany, which had hosted the pilots for the unarmed operations in 2000, was unwilling to allow operators to conduct an extrajudicial killing from German soil. Ultimately, it was an innovation that gave rise to modern remote drone warfare as we know it today.

Big Safari often works at staggering speed, and the Predator program was no exception. The contractor designed the remote-operations system over the course of a few weeks, and the whole unit was deployed less than a month after he finished. (Nicknamed the Man with Two Brains, the contractor is a revered — and feared — figure in Big Safari's tight-knit community. He keeps his identity a closely guarded secret; when I interviewed him for *Wired* magazine in 2015, I was not allowed to record our conversation, and before sitting down with me he scanned my person with a small black device to make sure I wasn't wearing a wire.)

Big Safari's ability to move so quickly stems largely from a development philosophy that engineers call "the 80 percent solution." To expedite deployment, Big Safari will work on an airplane until it's 80 percent done. The logic being that the final 20 percent of the work often takes just as long as the first 80 percent. In the Big Safari mindset, "good enough" is better than too late.

Because of its focus on speed and risk, Big Safari often chafes against the regular Air Force, which prefers its programs to be developed cautiously and meticulously, with each modification vetted by committees and program officers. During the Predator project, Colonel Bill Grimes, the commander of Big Safari at the time (and another Hall of Honor inductee), set up a secret phone line between his engineers in Ohio and test pilots in Nevada, allowing his staff to modify the aircraft without having to run each decision up the Air Force chain of command. The unit's informal motto is "Those who say it cannot be done should not get in the way of those doing it."

Unlike other military units, Big Safari doesn't hold open competitions for large contracts. Instead, it has special authority to grant deals to whichever company it chooses, at any price that it wants. One of Big Safari's regular contractors is Sierra Nevada Corporation, a privately owned company based in Reno, Nevada, that has received billions of dollars in contracts from the Air Force since 2000.

Shortly before the new surveillance project was given to Big Safari, Meermans had retired from the Hill and taken a position at Sierra Nevada. On his first day at the company, Meermans learned that his old employer, the Air Force, had awarded his new employer the contract to serve as primary contractor for the wide-area surveillance system that he had conceived. He claims that he didn't know his new job would put him right back at the helm of the initiative, but he appeared to be glad that it had.

In spite of its secretive nature, Big Safari has made a tradition of bestowing on its projects exuberant, sometimes even goofy, code names: Stray Goose, Cobra Ball, Sweet Sue, Honey Badger, and so on. "Wide-Area Airborne Surveillance System" was always just a temporary placeholder. So someone in the community — Meermans thinks it may have been a retired chief master sergeant — had given the proposed camera a new name, Gorgon Stare.

When he heard it for the first time, Meermans thought, *What the hell is that?* In the Greek mythological tradition, the Gorgons are three monsters from the underworld. According to the historian Stephen R. Wilk in *Medusa: Solving the Mystery of the Gorgon,* the mythological creature's "rigid, fixed, penetrating, unblinking stare" is its defining characteristic. In *The Iliad,* the Gorgon is described as having "burning eyes" and a "stark, transfixing horror." The most famous Gorgon, the snake-haired Medusa, is literally petrifying, as any person who sets their eyes upon her are instantly turned to stone.

When I was reporting for this book, the mere words "Gorgon Stare" had a similarly petrifying effect on nearly all of the military officials and industry employees I interviewed. On more than one occasion, the person I was speaking to simply walked away when I mentioned the program so as not to run the risk of accidentally divulging a classified detail. Even those who did stay and talk often hesitated to call it by its name.

In *The Odyssey*, Odysseus is seized by "cold fear" at the very thought that "dreadful Queen Persephone might send / the monster's head, the Gorgon, out of Hades." As it turned out, that was the whole idea. "Ultimately," Meermans said of the name, "it made a lot of sense."

DARPA Hard

In the interest of completing its projects as quickly as possible, Big Safari's new technologies are rarely developed in-house. Instead, engineers utilize existing components, a process they liken to picking athletes for a sports team. But when Congress conceived the program that became Gorgon Stare, no such camera—not to mention the kind of computer that would be required to power it—was available. Someone would have to will the technology into existence. As it happened, a special military lab already had something in the pipeline.

In the defense community, there is a term, "DARPA hard," that refers uniquely to the type of engineering that happens at the Defense Advanced Research Projects Agency, the storied military-development unit that tackles blue-sky science and engineering challenges that other government agencies might consider too daunting or risky. Things like hypersonic cruise missiles that fly at five times the speed of sound, programmable microorganisms, intelligent drone swarms, and so on.

Securing a position as one of the agency's program managers is also hard. When Brian Leininger set out, in 2006, just as Congress was secretly directing the Pentagon to build a new unified wide-area system, to apply for a position as a "PM," he understood that the only way to get the job was to pitch the agency's director, Anthony Tether, on a "DARPA hard" idea.

Leininger had worked at Lockheed Martin for 25 years, at one point overseeing the Maritime Systems & Sensors division, a $4-billion-a-year business. He was versed in silicon design, fiber optics, radar technology, and other emerging technologies. After hearing about Constant Hawk, Leininger decided to approach Tether with an idea for an enhanced wide-area-camera program. In an interview with the DARPA director, Leininger pointed out that the existing wide-area surveillance cameras were not sufficiently powerful to follow anything smaller than a vehicle. But networks, Leininger explained, are made up of people, not cars, and an effective camera ought to be able to track individuals on foot.

Leininger also believed that the cameras ought to cover a much wider area than the current cameras. Though the 30 square kilometers that Constant Hawk recorded was much larger than the coverage area of any other airborne video system in the Pentagon's arsenal, it was still not enough to map out the large, diffuse networks that were proving to be the most difficult to fight. Fifty percent of the vehicles of interest that analysts tracked in the footage left the camera's field of view before stopping at a location that ground troops could visit.

Capturing a wider view in enough detail to track individuals on the ground would require a much large number of pixels. Even a 100-megapixel camera (which did not yet exist) wouldn't be powerful enough. And because people on foot move in less predictable ways — unlike a car, a person can stop and turn around in the space of half a second — the camera would also need to have a faster frame rate.

A number of engineers and officials, including John Marion and Dan Cress, had urged DARPA over the years to adopt or create a wide-area-surveillance program of its own. It was an ideal technology for the agency — audacious, complex, and time-sensitive. But Tether had resisted their entreaties, perhaps because they didn't seem sufficiently DARPA hard.

Nobody could claim that Leininger's proposal wasn't hard, and in March 2006 he was hired. Leininger soon drew up a detailed, and daunting, proposal. His camera would fly at an altitude of 20,000 feet in a figure-of-eight orbit. It would cover 50 square kilometers at enough resolution and with enough frames per second to track even fast-moving individuals.

Such a camera would obviously create far too much information to send down to individual soldiers on the ground — that would be like trying to send 40,000 photographs via email to a friend every second. Instead, the camera would transmit only 65 soda straws of live footage — "chip-outs" — cut from the entire frame of view down to analysts on the ground, giving each a Predator-like view that they could independently steer to any area within the camera's field of vision. Leininger had also decided that the system needed to be small enough to fit on a drone.

When Leininger explained the technical requirements to his newly hired deputy, Richard Nichols, a former Army infantryman, Nichols thought he was "off his rocker." Nichols was not alone in his skepticism. In several meetings with John Marion, the Constant Hawk engineer had encouraged Leininger to curb his ambitions. A camera on the order that he was proposing would be a technological nightmare, Marion said. It would be massive, and it would need vast amounts of computing power, since a faster frame rate, in turn, requires faster computers and more onboard memory. But, in the manner of Big Safari, Leininger wouldn't let those who said it couldn't be done stand in the way of someone doing it. He listened to the arguments against, then ignored them.

DARPA approved the full proposal in November 2006. The agency called it the Autonomous Real-Time Ground Ubiquitous Surveillance Imaging System, ARGUS-IS for short. It is often referred to simply as ARGUS.

In Greek mythology, Argus is a fearsome giant who is tasked by the goddess Hera to guard her priestess, Io, who had been turned into a cow. Argus, who is also referred to as Panoptes (which means "all-seeing"), is particularly adept at this task thanks to the 100 eyes covering his head. Here's one typical description of the character, from the poet Elizabeth Browning's translation of *Prometheus Bound:*

The herdsman Argus, most immitigable
Of Wrath, did find me out, and track me out
With countless eyes, set staring at my steps.

ARGUS was therefore an appropriate name for what would be the most all-seeing camera ever built. Even after DARPA formally launched the effort, however, many officials still refused to accept that the proposed system, like its hundred-eyed namesake, was anything more than a figure of pure fantasy. When Leininger toured the Pentagon in search of a military unit that would be willing to put money behind the concept and take it to battle, some said they would only believe the technology was possible when they saw it in finished form. Others just shook their heads, unconvinced that the idea Leininger described was a serious proposition.

The All-Seeing Eye in Your Pocket

One of the more obvious challenges of Leininger's proposal was achieving the vast pixel count required for the camera. After building the computer for Constant Hawk, the surveillance-technology team at MIT's Lincoln Labs had been examining this very problem. Bill Ross, the team lead, knew that simply strapping more individual video cameras together wasn't an option, as this would add too much weight.

The heart of any digital camera is the chip, a flat panel of sensors that absorbs light entering through the lens and converts it into a picture, with each pixel representing the exact color and intensity of light that corresponds to its place in the frame being captured. For another program, the Lincoln Labs team had been experimenting with cell phone camera chips, which had shrunk dramatically in recent years thanks to intense competition among manufacturers seeking to build the smallest, lightest, and sleekest handheld devices. Using a five-millimeter-wide cell phone camera chip, they had managed to build a tiny surveillance device that could operate on a single battery for extended periods of time while producing high-resolution imagery. Ross was evasive about which government organization the device was built for. To me, it sounded like something you might hide inside a smoke detector in someone's living room.

It occurred to Ross and his colleagues that the same character-istics that made the chips ideal for small hidden spy cameras might also be useful for building a very large camera. Though each chip had only a few megapixels, they could be stitched together into a large mosaic that sat behind a single lens. A camera with enough chips could, in theory, produce much larger images than Constant Hawk or Angel Fire.

With funding from the Rapid Reaction Technology Office, the program that had partnered with the CIA to fund the initial devel-opment of the program that became Constant Hawk, Ross's team ar-ranged 176 of the five-millimeter cell phone camera chips into four identical grids, each behind a separate lens. The resulting camera could collect 880 million pixels per frame, 10 times more than Con-stant Hawk and 176 times more than a top-of-the-line cell phone. They called it the Multi-Aperture Sparse Imager Video System — MASIVS for short.

In 2007 Leininger visited Lincoln Labs to see MASIVS in devel-opment for himself. Though he was put off by the engineers' uncon-cern for protocol (they weren't wearing antistatic devices, which are used for protecting delicate electronics during manufacturing), their cell phone chips idea had potential. An order of magnitude increase in pixel count was a quantum leap as far as surveillance technology was concerned — and from what he could tell, the MIT team was well on its way to achieving it.

Taking Shape

In November 2007, just a few months after Angel Fire joined Con-stant Hawk on deployment in Iraq, DARPA awarded the defense giant BAE Systems $18.5 million to build a cell phone–based sur-veillance camera for ARGUS. BAE's effort would be led by Yiannis Antoniades, a Greek engineer with a bushy mustache, a thick nest of curly black hair, and a wry sense of humor. Leininger gave the com-pany two years to come back with a finished system.

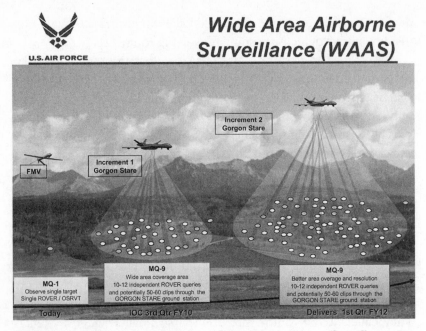

A slide from a 2010 presentation by a senior US Air Force officer illustrating the difference between soda-straw surveillance (far left) and the two planned Gorgon Stare systems. *US Air Force/David Deptula*

If ARGUS lived up to its original pitch, it would be exactly what Big Safari had been looking for. And so, in a deal largely brokered without congressional oversight, Air Force Secretary Norton Schwartz and Anthony Tether, DARPA's director, signed a memorandum of agreement in the fall of 2008 stating that DARPA would provide ARGUS cameras to the Gorgon Stare program as soon as they were ready.

As it began building the camera, BAE selected cell phone chips made by a company called Micron—an outfit that also produced chips for the iPhone. Thanks to the millions of consumers lining up to buy smartphones around the globe, the chips cost only about 15 dollars each. BAE proposed to stitch the chips into four panels, each behind an individual lens, a design similar to the Lincoln Labs camera. But BAE's design would utilize more chips. All told, the full camera would consist of 368 imagers, more than twice as many as MASIVS.

This left the problem of the computer, which would need to be capable of supercomputer levels of power to stitch together the 368 camera feeds, orient, and smooth out the imagery. The solution for that challenge also came from another unexpected corner of the commercial world: the video game industry.

Like wide-area cameras, computer games generate thousands of pixels that must be processed and assembled into coherent images. Over the preceding years, the computer gaming industry had seen an arms race to develop the most powerful graphics processing units (GPUs for short). By 2002, one firm, Nvidia, was doubling the capacity of its GPUs every six months, three times faster than posited by Moore's law, a theory that predicted—accurately—that the number of transistors that can be fit onto a single computer chip (and thus, by extension, the power of that chip) would double every 18 months.

To run its MASIVS video feed smoothly, MIT had built a processing unit using graphics chips from a PlayStation. Meanwhile, at Lawrence Livermore, a researcher named Sheila Vaidya, who had been leading a project that was exploring how to take advantage of recent advances in the video game industry for surveillance, built a similar unit using the insides of an Xbox.

As a result of outpacing Moore's law, these computers were much more powerful than those on Constant Hawk and Angel Fire, and yet also much smaller and lighter. The use of GPUs for wide-area aerial-surveillance processing systems soon became an industry norm. Many of the battlefield systems that operated in the ensuing decade and beyond share much of their computational DNA with your average video game console.

Drawing from these designs, BAE built a computer about the size of two shoeboxes packed with 33,000 processing elements. According to Antoniades, never before had so much processing power been packaged in a device small and sturdy enough to fit on a drone.

In early 2009, to get a sense of what the final system might be capable of, the engineers set up a single panel of cell phone chips outside a BAE facility in Acton, Massachusetts. With 450 pixels, this camera, technically just a quarter of a full ARGUS, would have already been the second-largest system in existence at the time. Scrolling around the test images, the engineers looked at a neighboring facility's car park, more than a football field's length away. Zooming in a little closer on one particular car, they realized they could read not just its number plate, but also every single visible license plate in the lot.

"We Did What We Said We Would Do"

BAE finished building the first camera, on schedule, in August 2009. In its final form, ARGUS had 1,854,296,064 pixels, enough imaging power to spot an object six inches wide from an altitude of 25,000 feet in a frame twice the width of Manhattan. It generated 27.8 gigabytes of raw pixel data, enough to fill six DVDs, *every second*. Downloading the raw data in real time would require an internet connection 16,000 times faster than the fastest wireless internet service available in the United States in 2017. Just processing all the pixels put the Xbox-styled computer's 33,000 processing elements through 70 trillion operations each second.

BAE began testing the camera aboard a Black Hawk helicopter over Fort A. P. Hill, an Army base in Virginia. With the Black Hawk hovering 10,700 feet above the ground, the engineering team tracked a green mowing tractor tracing lazy loops along a parched lawn on a remote corner of the facility. They then watched two people walking on a road. The footage was so clear that you could see the heat radiating from the ground and pick out the windshield wipers on the cars.

Two months later, the team conducted another set of flights at Marine Corps Base Quantico, also in Virginia. Yiannis Antoniades was watching the operation from a white tent in a corner of a parking lot. When the operators switched on the camera, 40 square kilometers of real estate encompassing the entire base and surrounding area, all the way across to the opposite bank of the Potomac River, came to life on the screens in front of him.

One of the staff members stepped out into the parking lot. In the video, the image was so sharp that you could make out the slight hunch in his shoulders, even though it was taken from five kilometers above his head. "I think we've done something good here," Antoniades said to himself. "We did what we said we would do."

Seven years later, I saw the same footage from the flights at A. P. Hill and Quantico, and even then it felt like I was peering into the future. At one point in the Quantico footage, two people meet in a parking lot. One hands a briefcase to the other, then they head their separate ways. The camera follows them both. The two could have walked three miles in any direction and the camera would still have had them squarely in view.

A screenshot from the Defense Advanced Research Project Agency's ARGUS-IS tests over Marine Corps Base Quantico in October 2009. The image, about one-ten-thousandth of the total area that the camera recorded from its orbit at 17,500 feet, shows an employee doing stretches in a parking lot. *US Defense Advanced Research Projects Agency/Brian Leininger*

The footage is riveting to watch; it's almost impossible not to project story lines onto the actions of the antlike figures on the screen. A car cuts diagonally across a parking lot — the driver must be in a rush, one thinks. Two people meet and speak briefly outside a building — a chance encounter, or is something illicit afoot? A pickup truck makes a sudden U-turn, seemingly for no reason — maybe the driver knows he's being watched and is employing countersurveillance techniques.

Shortly after the test flight in Quantico, DARPA invited dozens of Pentagon and intelligence officials to a demonstration at the base. On the day the demonstration was scheduled, poor weather prevented the agency from putting on a live flight — the idea had been to track each participant as they drove into the facility — and instead the team played the footage that I saw in 2016. But the effect was the same. According to several attendees' accounts, many officials found the reel just as mind-altering as I did.

"We Can See Everything"

The war in Afghanistan had again flared up, and IEDs and ambushes were now the leading cause of both coalition and civilian casualties in the conflict. In 2009 alone, there were more than 7,000 IED attacks. Given the crisis in Afghanistan, Big Safari did not want to wait for DARPA to finish producing a full series of ARGUS cameras. So it had pressed ahead with an earlier, less ambitious design know as Gorgon Stare I — an 80 percent solution — that could be deployed while BAE put ARGUS through production.

Gorgon Stare was by no means the largest Big Safari program in financial terms, but Colonel Ed Topps, who became the unit's commander in 2010, considered it to be one of his highest priorities. No other system the unit was working on, he told me, had the potential to make the same kind of impact on the lives of the soldiers who were dying on the ground. With the camera flying over the battlefield, Topps wrote to me in an email, "we could find the factory, the warehouse, the bomb making location and GET THAT INDIVIDUAL."

Given Big Safari's reputation, hopes had been so high for Gorgon Stare that the Marine Corps and the Air Force had even canceled an order for five additional Angel Fires in order to make way for an undisclosed number of Gorgon Stares. But when the system was unveiled, it struggled to hit all the high notes. The daylight and infrared cameras, which had been developed by ITT Corporation, had roughly the same pixel count as Livermore's prototype camera, the Sonoma, meaning that it lacked the resolution necessary for the more complex and sensitive operations that were a priority for the ground forces waiting to take delivery of the camera.

The cameras and their accompanying processors weighed 1,100

pounds, but the Air Force was insisting that the Gorgon Stare drones carry missiles in addition to the camera, and the effort to fit everything on a single airframe was adding time and expense. In response to a request for $79 million in additional funding for the program in 2009, the Senate Armed Services Committee recommended that the program be terminated instead.

In 2010 an Air Force test squadron assigned to evaluate Gorgon Stare's readiness for deployment found the system seriously lacking. It could transmit only small subsections of the full view of its camera from air to ground. It often skipped frames, which made tracking vehicles difficult. Its inability to follow people was noted as a major shortcoming. The ground stations could link to only one Gorgon Stare at a time, meaning that it was impossible to arrange the same 24/7 unblinking rotation that the Air Force employed with its regular drones. There was an average of 3.7 technical failures per flight.

"You're speeding," Topps was told by the testers, who summarized their experience in a leaked memo issued on December 30: "DO NOT field" unless these problems are fixed.

An Air Force spokesperson issued a response, noting that the leaked document was only a draft, and explaining that three of the issues that the test team discovered had already been resolved. Nevertheless, Air Combat Command was threatening to kill the initiative completely. By one estimate, the Air Force had by that point spent more than $500 million on development and research.

Perhaps sensing that the program was in need of a boost, a number of Air Force generals mounted a press offensive. Speaking to reporters from the *Washington Post*, they offered a dramatic assessment of the system's abilities. "There will be no way for the adversary to know what we're looking at," Major General Jim Poss is quoted saying, "and we can see everything."

(When I asked Poss about the article in 2016, half a decade after it was published, he seemed a little embarrassed, though he pointed out, somewhat proudly, that it had been quoted in a newspaper published by Hezbollah, the Lebanese political organization that is designated as a terror group by the US Department of State.)

A few weeks later, as the program continued to teeter on the verge of termination, General James Cartwright, the vice chairman of the Joint Chiefs of Staff, visited the Gorgon Stare unit as it was conducting final testing at Creech Air Force Base in Nevada. In a ground control station, Topps showed Cartwright the live surveillance footage from one of the Gorgon Stare–equipped Reapers flying overhead. "It's good enough," Topps recalls Cartwright saying. (Cartwright declined to confirm the anecdote.) So said, so done. In Topps's account, this was enough to keep the program alive.

The first four Gorgon Stare–equipped Reaper drones were deployed to Afghanistan three months later, in the spring of 2011. In the field, the system achieved mixed results. One engineer who reviewed some of the early footage called it "just terrible." But the Gorgon Stare drones could watch an area more than four kilometers wide, and beam ten individual chip-outs from the full image directly to units on the ground who could steer their view anywhere they wanted within the full field of view. As a result, the four Gorgon Stare Reapers were employed prodigiously; in the first three years in operations, the aircraft flew 10,000 hours of surveillance over Afghanistan.

Aircrews stationed in Nevada piloted the drones while teams in the field controlled the cameras. US Central Command focused the aircraft on "large population centers," though it still hasn't disclosed the exact areas of operations. Responsibility for imagery analysis was assigned to the 497th and 548th Intelligence, Surveillance, and Reconnaissance Groups, both divisions of the 480th ISR Wing (motto: *Non potestis latere.* Translation: "You can't hide"). After each flight, a group of analysts downloaded the footage and perused it for insurgent network activity, comparing it to footage from earlier missions to build a "God's eye view" of the insurgent networks. At any given time, the Air Force maintained a 30-day archive of data at the operating base. A more thorough analysis was then carried out at intelligence cells in Virginia and California.

Ed Topps traveled to Afghanistan shortly after Gorgon Stare was deployed and visited the operations base. He sat with an analyst who picked a random car driving down a city street and backtracked it to the house where it had come from. To Topps, it looked just like a movie being played in reverse.

Oculus semper vigilans

In 2012, the year after Gorgon Stare I was deployed, Big Safari took delivery of 10 complete ARGUS cameras from DARPA, and thanks to favorable feedback from commanders in the field, the program was allowed to progress unimpeded. The second iteration of Gorgon Stare was completed 18 months later. Nine drones were produced in total.

Most technical details about Gorgon Stare II remain closely guarded state secrets. What we do know is that each system consists of two long pods hung from the Reaper's wings in the spots where the bombs and the missiles would usually be. The pods house an ARGUS and a smaller wide-area infrared camera, developed by the contractor Exelis. An unpublished Air Force report from 2014 suggests that Gorgon Stare is also equipped with a signals-intelligence sensor that would allow operators to intercept adversary radio chatter and phone calls, presumably for use by the NSA, but nobody I asked would either confirm or deny this report.

Gorgon Stare II deployed to Afghanistan in 2014. A military patch for the program that was briefly posted on the Sierra Nevada website features a Medusa's head staring down on two small silhouetted figures. True to the original mythological figure, she has snakes for hair. Her teeth are long and sharp. One of her eyes is green, representing the daytime ARGUS camera; the other is red, for the infrared sensor. The patch bears the motto *Oculus semper vigilans:* "The eternally vigilant eye."

Like the technical specifics of the cameras themselves, all information regarding how Gorgon Stare is used in the field is technically classified. The unit operating Gorgon Stare II from Bagram Air Force Base held total reign over the country's cities. A single drone could take in 40 square kilometers, meaning that it could cover almost the entire city of Kandahar — and up to 100 square kilometers if it flew at higher altitude. It could probably beam chip-outs down to as many as 30 ground units at once. General Larry James, who ran the Air Force's intelligence, surveillance, and reconnaissance pro-

An MQ-9 Reaper equipped with Gorgon Stare departing for a surveillance operation from Kandahar Airfield, Afghanistan, in 2015. *The appearance of US Department of Defense (DOD) visual information does not imply or constitute DOD endorsement. US Air Force/Tech. Sgt. Robert Cloys*

grams in the lead-up to Gorgon Stare's deployment, describes this as "the synoptic view."

According to Colonel Mark Cooter, who served as commander of the unit responsible for analyzing Gorgon Stare imagery, teams working with the footage could often track a range of suspects while also keeping an eye on friendly forces at the same time. With just a couple of aircraft, one could effectively watch a single area endlessly.

The technology was not without its issues. For reasons that were never explained to DARPA's engineers, Sierra Nevada declined to use DARPA's supercomputer, and instead opted for a processor that captured only two black-and-white frames each second. Other sources, speaking off the record, grumbled about reliability issues.

Even so, the program won a number of admirers. John Marion, the engineer who had urged DARPA's Brian Leininger to be less ambitious in his approach, called it "astoundingly good." The Man with Two Brains, who would only say he was involved in the development of the system, called Gorgon Stare "an orthogonal expansion into a whole other dimension of surveillance."

It also soon began to be used in roles well beyond its original mission set. In addition to seeking out insurgents and IEDs, the camera's TiVo features were likely used to locate mortar teams attacking ground forces and bases, to survey sites in the lead-up to ground raids, and even to spot smugglers along the border (presumably with Pakistan). If analysts found an activity of interest, they could isolate the clip and push it up the chain of command via email in a matter of minutes. Mark Cooter, the intelligence commander, alluded to having worked on certain missions that the creators of the technology probably hadn't envisioned. In southern Afghanistan, he said, the line between counterinsurgency, counterterrorism, and counternarcotics was not always clear.

Within months of the initial deployment, intelligence cells working with the system asked for more aircraft, and the program was redesignated from its "Quick-Reaction Capability" status — a sprint effort — to an "Enduring Capability." The following year, in 2015, the Air Force deployed Gorgon Stare to Syria as part of the operation against ISIS.

In the years since, the Pentagon has invested tens of millions of dollars in the continued development of the system. Responding to an "urgent operational need," in 2015 the DOD allocated $10 million to develop a beyond-line-of-sight communication system that would enable Gorgon Stare to fly more than 500 miles from the operators, an effort that was ongoing as of 2018. The Air Force also worked for several years to integrate into the system a device called a Near Vertical Direction Finding sensor, which can locate a large number of communications devices within a surveilled area.

In 2017 the House of Representatives Committee on Armed Services described Gorgon Stare as "invaluable." The same congressional report cites "numerous" combat units that have referred to it as a "critical" system. Michael J. Kanaan, an intelligence official who was the only active service member the Air Force formally permitted to speak to me on the record about wide-area surveillance, said that demand for systems like Gorgon Stare will only increase in the future — though he, too, refused to call it by its name.

The Pantheon

Less than two decades since wide-area motion imagery began life as little more than a fixture of a Hollywood screenwriter's overactive imagination, it is safe to say that it is now very much here to stay.

Shortly after becoming involved with the early development projects, Dan Cress, the CIA official, began organizing yearly wide-area-airborne-surveillance conferences, where different groups would meet in the desert to discuss, demonstrate, and collaborate on their programs. Initially a small niche gathering, by the time Gorgon Stare was deployed, the conference drew over 100 companies, labs, and organizations working on a range of secret wide-area surveillance systems that, according to Marion, "can tell where you've been and what you've been eating."

In the intelligence community, all-seeing video became institutionalized. The days of couriers shuttling suitcases full of hard drives back from the battlefield are long since gone. The National Geospatial-Intelligence Agency, the spy agency responsible for aerial surveillance and reconnaissance activities, now channels all-seeing surveillance footage directly to its analysts in the United States. A program called Hiper Stare, developed by the PIXIA Corporation, allows analysts to access the government's archive of wide-area video from anywhere in the world. (Rahul Thakkar, one of the engineers who developed the software, received an Oscar for his work on the computer graphics for a number of popular films, including *Shrek*.)

Other wide-area systems were developed in parallel to Gorgon Stare and they, too, have been used extensively. In 2007 the CIA and the NSA, which was in the midst of an ambitious effort known as Real Time Regional Gateway that sought to vacuum up every Iraqi electronic communication that might be relevant to an IED cell, sponsored the development of a system that combined a wide-area surveillance camera with a number of the NSA's classified electronic surveillance sensors. After the airplane performed impressively at a large military and intelligence exercise called Trident Specter, the Air Force adopted the idea and mounted the system on an innocuous twin-prop plane, which it christened Blue Devil.

Blue Devil went to Afghanistan in December 2010, just 280 days after the Air Force formally began working on the program. Shortly thereafter, it was joined by a new variant of the Army's Constant Hawk equipped with MIT's 800-megapixel MASIVS.

Concurrently, the Army contracted Logos Technologies, reportedly with CIA backing, to put wide-area cameras on tethered blimps to help protect forward operating bases in Afghanistan from Taliban mortar attacks and ground assaults. In a matter of weeks following the contract, the company sketched a design for what would go on to become the most widely used wide-area camera developed to date, the 440-megapixel Kestrel.

First deployed in August 2011, the Army's 38 Kestrel surveillance blimps accrued over 200,000 flight hours. As of 2018, they remain in use at two bases in Afghanistan and one in Iraq. Although Kestrel is primarily used for defensive surveillance, it has also been used as an offensive intelligence-collection device to identify suspected insurgents. In one operation described to me, US analysts used the system to watch the funeral of a suspected Taliban leader, and then tracked the attendees back to their homes.

At DARPA, Brian Leininger, the ARGUS program manager, led an effort to produce an infrared camera with the same coverage area and resolution as the daylight version: ARGUS-IR. Though the materials and cooling systems necessary for the design could at best be described as experimental technologies at the time, Leininger's former employer, Lockheed Martin, completed the effort in 2014, producing two cameras of astounding complexity and, judging by the snippets of footage I reviewed, from a flight over Quantico, remarkable surveillant power.

As with the original ARGUS, it didn't take long to find a customer willing to take the infrared version to battle. Joint Special Operations Command, the Pentagon unit responsible for sensitive counterterrorism operations, including drone strikes and raids against high-value targets, soon began building a surveillance aircraft equipped with ARGUS-IS and, according to sources, the ARGUS-IR. Once deployed, the airplane will be able to stare out over entire cities with equal acuity at high noon and midnight.

To be sure, not all WAMI programs have been successful. In

2012 the Army attempted to mount ARGUS on an experimental un-manned helicopter called the A-160 Hummingbird, but abandoned the project after it crashed during a flight test days before it was scheduled to deploy. The next year, the Army called off an even more preposterous program that aimed to fit a seven-story-tall blimp with a wide-area-surveillance system. The Pentagon was forced to sell the prototype aircraft back to the manufacturer for $301,000, one-one-thousandth of the $297 million that it had spent developing it. It was therefore no surprise when the Air Force's attempt to build a second, larger and more complex version of Blue Devil on a similarly gigantic blimp turned out to be an equally expensive disaster.

But these costly missteps appear to have done little to dim the Pentagon's appetite for wide-area surveillance. In 2018 the Air Force's 427th Special Operations Squadron, a tiny unit so secretive that the Pentagon is reluctant to even acknowledge its existence, deployed modified CN-235 cargo planes—equipped with all-seeing surveil-lance cameras mounted on the left side of the fuselage—to Syria, where they shared the sky with Gorgon Stare. The Air Force will say only that the 427th specializes in infiltration and exfiltration. But the unit is also known to work closely with the CIA.

The Army, meanwhile, is building a fleet of surveillance planes known as the Enhanced Medium Altitude Reconnaissance and Sur-veillance System, for use largely in special-operations missions out-side of declared war zones, some of which will be equipped with wide-area sensors. As of 2017 the Army had already deployed four of the aircraft for use in Africa and South America, though it has of-fered scant detail as to what, exactly, they are being used for.

The Army is also readying a fleet of multisensor spy planes called the RO-6A, which will be equipped with a 174-megapixel six-lens camera capable of reading the number on the back of a football play-er's jersey, and those of all his teammates, from four kilometers above the earth. They will be ready for deployment by 2020.

In 2017 the Department of Defense also began building a new all-seeing system destined for the Army Special Operations Command called simply "Advanced Wide Area Motion Imagery," which will be smaller and lighter than existing cameras but just as powerful; DOD is also exploring a set of technologies that suggest it is looking to build a wide-area sensor to be used aboard the Army's version of the Predator drone, the Gray Eagle.

The Air Force has even begun exploring the idea of a replacement for Gorgon Stare. An internal study on the topic from 2018 noted 11 types of targets the Air Force wants to track more effectively with this new system, including forces in densely forested areas and individuals on foot in urban settings.

And these are only the projects that are known. Sources I spoke to from Big Safari, Livermore, the CIA, the Army, and the Air Force all alluded to other systems, the details of which they said they were unauthorized to discuss.

Unanswered Questions

The story of the all-seeing eye is remarkable. But it is incomplete. While it is clear why the defense and intelligence establishment set out to build these enormous cameras in the first place — to stop IEDs, to attack terrorist networks, to save lives. But no substantive information has been made public about what happened when these systems were actually put to their intended mission. Did they work?

Many of those involved, like Big Safari's Ed Topps, claim that Gorgon Stare, like the other systems, has saved lives. But nobody is willing to divulge exactly how it did so. The furious and ongoing investment in the technology seems to suggest that it had some kind of effect, but the Air Force declined repeated requests for even an approximate indication of WAMI's impact on the battlefield. These details are considered classified operational information. Until they are declassified, it remains to be seen whether the all-seeing eye lived up to the noble aspirations that spurred its creation.

The little evidence that is available suggests that WAMI's short history has been a violent one. A laudatory webpage for an Air Force engineer who won the coveted DOD "Scientist of the Quarter" award in the fall of 2014 noted that the Air Force's Blue Devil, which he had worked on, was "credited" with the capture or killing of over 1,200 people. An astonishing toll on its own, the figure is all the more chilling when you consider that the Blue Devil fleet consisted of just four aircraft that had been operational for only a little over three years when the page was posted.

According to one Air Force official, Blue Devil helped lead forces to "a number" of high-value targets involved in the IED networks, but it's likely that the majority of those 1,200 individuals rounded up with the system were little more than low-level operatives. The US air campaign in Syria, which is directly supported by Gorgon Stare Reapers and the 427th's CN-235s, has likewise decimated ISIS's leadership and rank and file alike, but the civilian toll of that campaign has been heavy. If WAMI has had any part in the operations that took the lives of either those combatants or civilians, or in missions that have directly saved the lives of others, the Pentagon won't say.

Either way, the public has a right to know, not just so that we can assess the merits or demerits of the all-seeing eye as a weapon of war but also, just as importantly, so that we can fully grapple with the prospect of its use as a tool in times of peace. An inside joke among the engineers who built the original all-seeing eye was that they might go down in history as the creators of Big Brother. Today, it doesn't seem all that funny anymore.

PART II

Our Overwatched Future

4

A Murder in Baltimore

IN THE EARLY hours of August 17, 2014, I was riding my bike along Lafayette Avenue in Brooklyn after a long night out with my brother and his wife. At 2:43 a.m., as I crossed Lewis Avenue, I heard a popping sound to my right and saw a group of four people darting silently into the darkness, leaving behind a fifth person who fell, first to his knees, then to his chest, in the middle of the street.

It took a few moments for me to realize that what I had just witnessed was a shooting. As I circled back around the block to the victim, I heard sirens approaching. A young man was lying on the ground, not moving. A few feet away, a teenage girl was standing with her hands over her mouth, staring at the boy with wide, teary eyes. The attackers were long gone.

Two days later, I managed to reach a detective who was working on the case, and he asked me to come to the local precinct to answer some questions. The victim, whose name was Taekwon Hart, was in critical condition. He was nineteen years old.

In an interrogation room, the detective drew a map of the intersection and some of the surrounding streets and asked me to show him how the assailants had fled. I took the map, rotated it north-up, and drew an arrow pointing west down Van Buren. I had no further information. I didn't know what the assailants were wearing or what they looked like.

A few days later, I called the detective to ask how Hart was doing. He was going to survive, the detective said, but the investigation had stalled. The shooting had joined the thousands of violent crime cases that remain open in New York City every year.

If an all-seeing camera like Gorgon Stare had been flying over

Brooklyn the morning of the shooting, Hart's assailants might not have walked free. In the minutes after the attack, police could have tracked the perpetrators forward in time as they escaped, as well as backward to where they had come from. Even if the police couldn't catch up to them that very day, they would nevertheless have had a set of crucial leads. The simple logic of wide-area surveillance — the more information you collect, the more likely you are to find your adversary — applies as equally to the home front as it does to the battlefield.

Like the US military, law enforcement agencies operate a variety of surveillance aircraft, but they have many of the same drawbacks as traditional military spy planes. The soda-straw sensors on most police helicopters, for example, are great for chasing a single Ford Bronco down a Los Angeles freeway. But they are incapable of watching more than one incident simultaneously.

And since police helicopters are extremely expensive — a single system may cost as much as $10 million to buy and thousands of dollars per hour to operate — even large departments can afford only a small fleet of aircraft, which are reserved for high-priority crimes. Most police work therefore takes place without an eye in the sky. A single WAMI system, on the other hand, could chase a Bronco *and* watch over a SWAT raid across town while also keeping track of a number of lower-priority incidents, all at the same time.

Nor are law enforcement aircraft particularly useful for the type of after-action investigative work that might have helped find Taekwon Hart's attackers. The New York City Police Department's helicopters usually take at least 10 minutes to arrive on-scene following a call for service. If a helicopter had responded on the night of the shooting — and as far as I can tell, none did — the shooters would have been long gone by the time it showed up. A WAMI would have been watching the whole time.

This synoptic view can also be useful in large, complex security operations of the type that might be necessary during a terrorist attack. Following a vehicle-ramming attack, say, analysts would be able to quickly backtrack the vehicle to its point of origin in order to establish whether the perpetrators are tied to a larger group that is planning further attacks the same day. In the aftermath of an attack,

analysts could likewise map out the choke points and assist respond-
ers in directing the flow of people away from danger.

All of the labs and companies that have produced WAMI were
aware of its potential for domestic use from the very beginning. In
the fall of 2002, just months after building the first wide-area prototype, the Livermore team attempted to assemble an airship equipped
with the system to send to Washington, DC, to help find the perpetrators of a series of sniper attacks that were terrorizing the city and
surrounding areas.

Ultimately, the DC snipers were caught before the aircraft was
ready. But the team was undeterred. A report that it produced in
2006 describes how Sonoma, the camera that would later evolve into
Constant Hawk, could track 8,000 targets simultaneously, making it
perfect for the "more ubiquitous and persistent surveillance" that, it
noted, was now needed in the post-9/11 era.

In the ensuing years, dozens of defense and security contractors
have worked to put these systems directly into the hands of US law
enforcement agencies at every level of government. As the wars in
Iraq and Afghanistan have wound down and demand from the Pentagon has slowed, these efforts have accelerated.

Sierra Nevada, the contractor that built Gorgon Stare, has pitched
its peacetime version of the camera, Vigilant Stare, to the FBI and
the Secret Service, among other federal agencies. BAE demonstrated
ARGUS for US Customs and Border Protection in June 2015, as well
as a number of other agencies that no BAE or US officials will disclose by name, and it markets its 100-megapixel Airborne Wide-Area
Persistent Surveillance System specifically for "harbor security, large
sporting events, and anywhere wide area surveillance is needed to
help protect life and commodities."

The defense firm Harris, which acquired the company that made
the infrared sensor for Gorgon Stare and more recently unveiled a
wide-area camera called CorvusEye 1500, has been particularly active in its marketing efforts. In August 2015 the company provided
one of its WAMI aircraft for Urban Shield, a large yearly law enforcement and emergency-preparedness exercise held in the San Francisco Bay Area. Responding to a fictional terrorist plot to sabotage
the local water supply, the airplane scanned a vast expanse of sub-

urbs, tracking both the "terrorists" milling about near the Dunsmuir Reservoir and the movements of the response teams closing in a few miles away.

Following the event, Harris lobbied the local sheriff who hosted the event for an introduction to officials in neighboring Santa Clara, which was preparing to host the 2016 Super Bowl (the effort was unsuccessful).

Logos Technologies, the Virginia-based contractor that makes a range of wide-area cameras, puts on marketing flights for domestic agencies every few months, while Commuter Air Technology, an Oklahoma-based contractor that operates a variety of special-ops aircraft for the Pentagon, maintains an airplane equipped with a 300-megapixel WAMI camera on constant alert at one of its domestic facilities. It claims that it can respond to emergencies anywhere in the continental United States within 24 hours.

MAG Aerospace, an aerial-imaging and -surveillance services firm that has, according to promotional materials, surveilled over 13 million square miles of territory for military units, offers a Cessna

A civilian GA8 single-prop plane equipped with a Consolidated Resource Imaging WAMI surveillance system. *Consolidated Resource Imaging/Nathan Crawford*

equipped with a state-of-the-art Logos camera that can likewise be deployed on short notice for customers anywhere in the United States. Consolidated Resources Imaging, the company that ran Constant Hawk operations in Iraq, provides a similar set of services with a GA8 single-prop airplane. Like all domestic WAMI aircraft, to the untrained eye CRI's GA8 has few external markers to differentiate it from a regular benign civilian aircraft.

The defense contractor L-3 sells a specialized intelligence airplane, SPYDR, that can be configured to carry wide-area-surveillance systems, specifically certified for use in domestic airspace. The company's tagline for the system: "Your target cannot escape SPYDR." (In the fall of 2018, L-3 announced a merger with Harris, creating the world's sixth-largest defense company—one with an extensive WAMI repertoire.)

Much of the marketing material for these campaigns is available online, and it is eye-popping stuff. In a promotional video for CorvusEye, a stern narrator explains how a camera that watches an entire city is perfect for fighting "illegal and threatening activities around borders, high-profile events, and critical infrastructure." Logos's website features a sleek and cheery "What is WAMI?" primer. Northrop Grumman's promotional video for its 200-megapixel HawkEye features a heavily armed police force raiding a building under the airplane's watchful gaze. The clip has all the cinematic polish of a high-budget action movie.

Other companies that appear to be actively marketing and selling wide-area spy technology for domestic surveillance services include Special Operations Solutions, Stevens Aviation, Avcon Industries, Valair Aviation, Support Systems Association, as well as Panopses, a firm founded by Yiannis Antoniades, the engineer who designed ARGUS (Antoniades declined to provide much detail when we spoke, though perhaps the company tagline says it all: "The ubiquitous WAMI project").

These firms are determined to get the all-seeing eye into as many law enforcement agencies as possible, and their eventual success is virtually certain. Someday, every major city in the United States will be watched by a wide-area surveillance system. The question is, at what cost?

Man on a Mission

If there is one person who might be called the Henry Ford of wide-area surveillance, it's retired Air Force colonel Ross McNutt. Tall and imposing in build but avuncular in affect, McNutt did not invent the technology, but he has done more than any other individual to bring it into the mainstream. Like his colleagues on the Angel Fire and Constant Hawk projects, McNutt was enthralled by the idea of wide-area surveillance. While still in the Air Force, he had developed an audacious plan for several more development cycles of Angel Fire, culminating with a 24-lens camera with a complement of precision bombs to strike targets identified in the footage.

Like the others, McNutt always knew that WAMI's potential extended well beyond the battlefield to the domestic sphere. Shortly after leaving the Air Force in 2007 — following the transfer of the Angel Fire project to the Air Force Research Laboratory — he built an eight-lens camera modeled on Angel Fire and founded a company, Persistent Surveillance Systems, with the aim of providing wide-area surveillance services to domestic law enforcement agencies. In a matter of years, McNutt would find himself running the most extensive domestic wide-area surveillance program ever carried out in the United States — and generating a huge amount of attention, not all of it favorable, for both himself and the technology that he so adamantly believes holds the answer to many of the problems that beset the modern American city.

Upon founding PSS, McNutt sketched out an operating concept for domestic wide-area surveillance that, like the camera itself, mirrored the military's approach abroad. A small civilian airplane would trace high-altitude orbits over a city, feeding the video from the onboard camera in real time down to trained analysts in a command center on the ground.

If any crime occurred that might benefit from an aerial perspective, the analysts, who would be monitoring incoming emergency calls for service, could zoom in to the location and begin tracking anybody who had been involved in the incident both forward and backward. By the time detectives began working on the case, the company would have prepared a detailed map showing where suspects had been in the hours before and after an incident, who they had interacted with, and, ideally, where they lived.

It wasn't long before McNutt began to attract interest. His first customer was the Philadelphia Police Department, which contracted the company to carry out several trial flights in February 2008. Three months later, McNutt demonstrated the system for the Secret Service and the FBI with a series of live surveillance flights over Baltimore.

That fall, McNutt traveled to the International Association of Chiefs of Police annual conference, in San Diego, where he met José Reyes Estrada Ferriz, the municipal president of Ciudad Juárez in Mexico. Juárez had become an epicenter of the Mexican drug wars. By some estimates, residents of the city were statistically more likely to die in violence than civilians in Afghanistan or Iraq. Ferriz felt that Persistent Surveillance Systems was exactly what Juárez needed, and he hired the company to establish a full-time surveillance program.

Less than two hours into the company's first operational flight over Juárez, in early 2009, the camera recorded a man being shot in the head in an alleyway. According to McNutt, when police arrived on the scene, not a single witness was willing to speak about what had happened, even though PSS analysts had observed at least half a dozen people with a direct line of sight of the shooting. One individual, who appeared to have been the triggerman, had fled the scene in a car that had been idling nearby. The car had then taken a long, circuitous escape route across the city, stopping at three separate locations along the way.

PSS also tracked a second vehicle that had circled the scene of the crime twice before speeding off to a house on the other side of town, as well as a third vehicle that had fled from the scene moments after the shooting. Using Google Maps, the analysts were able to provide investigators with a list of addresses where each of the cars had stopped over the following hours.

Soon, the analysts were inundated with cases. After a different murder weeks later, the team found that the two likely assailants had interacted with the driver of a vehicle that had proceeded to head down to the local precinct. The team realized that it was a detective who was likely collaborating with the assailants. "It made you go 'Oh, shit,'" said McNutt as he later walked me through the footage.

Fearing that one of the cartels in the city might try to disrupt the surveillance operations, McNutt and his employer had taken measures to minimize leaks. Only a handful of officials in the Mexican government were informed about the program, and the company often fed its findings to a single law enforcement contact. McNutt also covertly provided intelligence to the US Drug Enforcement Administration and other US agencies, ostensibly as a form of insurance. The hotel that housed the analyst cell was kept under tight security, and the team was instructed to venture outside as little as possible.

In total, 2,643 people were killed in Juárez that year. McNutt showed me a list of crimes his team had observed and investigated during the run of the operation: "body dumping by a bridge; attempted execution; execution of a city official; execution on street; double murder; murder in a mall parking lot." He showed me one especially troubling clip in which a police officer was shot to death while driving to work. In the footage, the assailants display the cool, coordinated precision of a Navy SEAL team.

The PSS program was cut short by an abrupt turnover in city government just six months after it began. Even so, in that time the company's analysts had provided investigators leads for more than 30 murders.

A Billionaire's Bet

McNutt wasn't in Juárez long enough to significantly reduce the city's murder rate, but it did get him noticed back in the United States. In the fall of 2015 McNutt was contacted by representatives of John D. Arnold, a media-shy billionaire who had founded the energy-based Texas hedge fund Centaurus Advisors. Through his personal foundation, the John and Laura Arnold Foundation, Arnold had been funding the development of a number of new crime-fighting technologies, including a criminal-sentencing computer program called Public Safety Assessment that calculates an offender's risk of recidivism based on dozens of data points. Arnold had learned about PSS through a podcast, and he was interested in sponsoring McNutt's efforts.

For McNutt, who still hadn't managed to establish a long-term contract with a US city, it was a welcome offer. In the lags between law enforcement jobs, PSS had worked on less flashy projects, like surveying corn crops for Monsanto, but McNutt had remained fixated on the idea of getting the technology into police hands. He had come close to making deals with Los Angeles and Dayton, but Los Angeles had ultimately decided that the technology did not match the city's specific needs at the time, while Dayton's proposed program was called off after it was met with opposition from local residents.

A representative from Arnold's foundation asked McNutt to propose an American city where deployment of an extensive surveillance program might serve to establish the technology's credentials. McNutt wrote back 10 minutes later and suggested Baltimore.

In many ways, Baltimore was a logical choice. The city was battling a crime wave of warlike proportions. The previous year had been the bloodiest on record, and 2015 was on track to be even worse. Sixty percent of the city's recent murders remained unsolved.

The Baltimore Police Department had already demonstrated that it wasn't averse to new, experimental, even controversial technologies. The BPD had also recently completed the installation of a large and sophisticated network of CCTV cameras called CitiWatch that appeared to have led to a consistent drop in crime rates around the city. Since 2007, it had employed StingRay cell site simulators, which are used to track a suspect's cell phone, in more than 4,000 operations.

The Baltimore police had also been an early adopter of Geofeedia, a software program developed with funding from the CIA's venture-capital fund, In-Q-Tel, that monitored thousands of local social media accounts to help investigators find leads on past crimes, as well as identify people who, based on the content of their posts and their known social connections, could be deemed likely to commit future offenses.

In 2015 BPD used Geofeedia to assemble social media images from the protests following the death of Freddie Gray, a 25-year-old black man who sustained severe injuries while in its custody. Analysts ran the images through a facial-recognition system to identify protesters with outstanding warrants so that they could be arrested, according to a Geofeedia report, "directly from the crowd." After this and similar counterprotest operations in other US cities were revealed the following year, the major social media platforms revoked Geofeedia's access to their data and BPD terminated the program.

The city also appeared to be open to accepting offers of free, untested surveillance services. During the Freddie Gray protests, BPD had agreed to deploy software newly developed by a local social media monitoring company, ZeroFox, which claimed that it could identify "threat actors"—those individuals among the protesters who were most likely to incur or incite violence. Working pro bono, ZeroFox designated as threat actors numerous peaceful activists, including two prominent Black Lives Matter organizers whose actions a judge later found to be wholly protected by the First Amendment.

For years, McNutt had been courting the director of Baltimore's CCTV program, who maintained a keen interest in his work but had never been able to secure funding to hire the company. After hearing from the Arnold Foundation, McNutt contacted the official and explained that the money would be put up on the city's behalf.

Now the proposal quickly ascended the BPD chain of command. Within a month, Baltimore police commissioner Kevin Davis approved the program, and John Arnold made a personal donation of $360,000 to the city. (An Arnold Foundation spokesperson declined to comment for this book.) By January 2016, PSS was airborne over Baltimore. At first, the initiative, which had been given the intentionally anodyne name Community Support Program, was billed as a pilot project that would conclude in the fall of 2016, at which point it would be presented to the city government—which had not been privy to the deal—for approval as a permanent operation.

Under the terms of its agreement with the police, as often as the weather permitted the company would fly a single-prop Cessna with no police markings equipped with a 192-megapixel camera in a circular orbit from 11:00 a.m. to 8:00 p.m., with a 30-minute break to refuel. BPD had designated two orbits—one that covered more of the western portions of the city and another positioned slightly farther to the northeast—and would direct the company to fly one or the other based on where criminal activity was expected to be concentrated on any given day. The camera would generate one 32-square-mile frame per second. In a single flight, it might capture a homicide, three or four assaults, dozens of muggings, and hundreds of drug deals. Typically operated at about 10,000 feet above ground level, the aircraft would be difficult to detect with the naked eye. McNutt was instructed to focus on what he calls "unsolvable crimes," major violent incidents with no human witnesses.

The Operation

When I first spoke by phone with Ross McNutt in the spring of 2016, his tone suggested the controlled, slow-burning intensity of a military commander in battle. "I'm dealing with a murder, a mugging, and about forty dirt bikes," he said, "but what can I do for you, sir?" After I explained that I was writing a book about wide-area surveillance, he said that, as luck would have it, he was in the middle of conducting a surveillance program. If I promised to keep the details of what he was doing, and where he was doing it, a secret, and as long as the book wasn't published before the fall, I could come down and see exactly what he was up to.

I arrived in Baltimore on a muggy, overcast morning in mid-July. McNutt met me in a bland office building downtown and led me, through a heavy metal door that was unmarked but for a small sheet of paper with the words COMMUNITY SUPPORT PROGRAM taped to it, into a rather dreary office. In a large conference room, McNutt switched on a pair of projectors and typed some commands on a keyboard. A few moments later, the entire wall in front of us lit up with an aerial view of the city stretching from downtown Baltimore in the east to the Edmondson Village neighborhood in the west.

This was footage from the previous day; as the city was now cloaked in a heavy cover of low-lying clouds, the plane was grounded, giving the analysts a window to catch up on their extensive caseload.

McNutt began clicking a mouse. With each click, the footage moved forward by a single frame. Tens of thousands of cars were visible, hopping along the city's streets and avenues in one-second intervals. McNutt zoomed in. Tiny, single-pixel shadows seemed to be floating along the sidewalks. Those, McNutt explained, were people.

As McNutt and I conversed, a staffer walked into the conference room and hovered at the door. McNutt turned to him. "Yes, sir?"

"It's a homicide," the man replied.

"I should go deal with that," McNutt said, leaving me alone in front of his godlike view of Baltimore. When he returned, he appeared unfazed by whatever had taken place in the adjoining room.

Rather than hiring ex-military intelligence analysts, McNutt explained, he prefers to employ young civilians with experience in video games. The PSS software even uses familiar gaming keystrokes for certain commands. A good analyst can track a car three to four times faster than the car itself travels on the ground. In one case, analysts had followed a motorcycle allegedly involved in an assault on a police officer for over an hour as it threaded its way through the city. Within two hours, the suspects had been arrested. In some cases, analysts had managed to track a vehicle for up to four hours.

The analysts found that vehicles that had been involved in violent crimes tended to drive into the city's downtown, perhaps in the hopes of losing the police helicopters between the skyscrapers. After one homicide, a single vehicle that the analysts were tracking stopped at 22 different locations. A detective involved in one of the cases said that it would have taken weeks to piece together such a detailed narrative of the suspects' moves from CCTV footage alone.

In another instance, after tracking a suspect from the scene of a shooting to a nearby townhouse, investigators found that the address had been the site of multiple calls for service in recent years. They scoured the names of men who had been involved in those incidents and checked their mugshots against CCTV footage of the suspect. Within minutes, they had a match. The suspect, who just a few minutes earlier had been nothing more than a fleeting pixel, now had a name. He was arrested that same day. (A BPD spokesperson did not respond to multiple requests for comment.)

Sometimes the most valuable information about a suspect is gleaned not from chasing them forward in time as they escape the scene of the crime, but rather backward to the minutes and hours leading up to the act. As the Angel Fire engineer Steve Suddarth, who has also been working to bring WAMI technology into domestic use, put it, "Bank robbers put a lot of effort into planning their escape and their hideout after their crime. Relatively few, however, take the same precautions before the action."

A suspect's movements prior to a crime might reveal their regular home address and help lead investigators to accomplices and associates. Even mundane information about a suspect's daily routine can be useful. In Juárez, investigators would use information about a suspect's whereabouts in the lead-up to a crime to dupe him into believing that an associate was cooperating with investigators. Thinking that his accomplices had turned against him, a suspect might then offer a confession in exchange for a more lenient charge.

Following a violent crime, the technology also allowed the analysts in Baltimore to identify bystanders who might have information about the incident but had not come forward on their own. The analysts were often just as dogged in their pursuit of witnesses as they were in chasing the perpetrators themselves. After one shooting that injured two elderly bystanders, McNutt's team tracked two dozen vehicles that had been in the area at the time in the hope of finding someone who would talk.

Persistent Surveillance Systems had also developed techniques for persistent surveillance that mirror those employed by the likes of the CIA in operations abroad. In particular, by offering a single, seamless view of an area, the cameras enabled the company to thread together complex narratives that might not have been easy to discern from ground level. The city offered many opportunities to demonstrate those capabilities. After several months of operations in Baltimore, McNutt's analysts noticed that certain vehicles of interest were moving around the city in coordinated patterns that reminded them of the tactics employed by Mexican drug cartels in Juárez.

At the time of my visit, the team was staking out one location in particular, a small, rundown convenience store that they believed was implicated in a large-scale criminal enterprise. Every day more than a dozen cars would visit the store, in some cases coming straight off a nearby highway just to make a two-minute stop. To the analysts, it seemed unlikely that all these people were driving so far out of their way just to buy a soda. Some of the vehicles that came by the store had later been connected to shootings and stabbings.

Technically, the Baltimore Police Department's CCTV data-retention policy required the city to discard any footage that wasn't directly related to an open investigation after 45 days. The policy might prove to be an obstacle to such an operation (which was not a formal investigation). But McNutt told me that because the camera will invariably capture evidence that is related to an ongoing investigation *somewhere* in the frame, the department could always find a reason to hold on to all the data indefinitely. (By the end of the month following my visit, the company had conducted at least seven such special operations and investigations.)

McNutt believed that the convenience store was selling ingredients used to cut heroin for resale on the street. If the police launched a raid, he was sure they could take down at least 15 distributors in one fell swoop. A police supervisor had approved the aerial stakeout, but there was no warrant, nor did the operation appear to be tied to an ongoing effort by detectives on the ground. McNutt was going it alone.

Above the Law

Nothing in US federal law prohibits the domestic aerial spying activities I have just described. When WAMI operators head out to fly military-grade surveillance cameras over populous areas, they might inform authorities, as Steve Suddarth did when I flew with him in Albuquerque, that they are conducting a "photo mission." But even that seemingly benign disclosure isn't necessary from a legal standpoint.

How so? Say you're on a commercial flight and you pass over Baltimore. You pull out your phone and take a picture. Much of the area that you have photographed is private property, but have you violated anybody's privacy? You'd probably say no, and you'd be right.

But what if, instead of your phone, you use a professional camera equipped with a telescopic lens that's strong enough to make out individual people in their backyards? Though your actions might raise some eyebrows among your fellow passengers, they are, from a legal standpoint, no different from the first example.

Let's say you take it a step further. You fly a helicopter over the city at a thousand feet. Now, with your telescopic camera, you can even make out distinctive features of the people in your frame. Surely this isn't legal, you might say. Surely a bright line exists between snapping a photo with your phone from an airplane window and focusing a telescopic lens from a few hundred feet over someone's backyard.

But there isn't. Even if you happen to record someone sunbathing in a backyard, it is not your fault. It's theirs, for not taking better precautions to protect themselves from aerial observation.*

This is because the airspace over America falls into the same legal category as other public spaces like sidewalks, roads, parks, and beaches — and it isn't illegal to take photographs of private property, or private citizens, from public space. As such, we have no expectation of privacy from above. (Publishing particularly sensitive images *may* land you in trouble, but here we are talking only about the act of capturing that data.)

Even the most secretive organizations in the US government are not necessarily safe from aerial observation. In the spring of 1998 an NSA employee who goes by the nickname Cheebie was standing in the parking lot of the agency's headquarters in Maryland when a civilian helicopter equipped with a strange round object under its nose — a camera, he realized in horror — flew directly overhead. A group of employees inside the building worried that the helicopter crew might have even been able to see into their offices through their open blinds.

As it turned out, the helicopter was being operated by a film crew working on, believe it or not, *Enemy of the State.* They were taking a series of establishing shots of the agency's sleek glass complex. *Wasn't there anything the agency could have done to prevent the flight?* Cheebie and other employees asked the NSA administration. There wasn't. "Believe me," wrote one public relations official in response to the queries, "we tried."

---

* To be sure, if they do take precautions, and you intentionally find a way to circumvent those precautions, then, yes, you are violating their privacy.

A few days earlier, members of the film's production team had met with a group of NSA officials, and had explained that they were planning a series of flights over the facility in the following weeks. The officials protested the idea, but they had no grounds, legally speaking, to stand in the way. The airspace over the complex is public. All the *Enemy of the State* crew had to do was file a flight plan and call it a photo mission.

Capitalizing on this gap, as some might call it, in standing privacy law, wide-area-camera manufacturers and users often turn the all-seeing eye on peacetime populations in the United States and elsewhere without their knowledge. Sierra Nevada flies Gorgon Stare cameras over suburbs in Colorado for testing. PV Labs, a Canadian firm, has flown its 300-megapixel camera over various US and Canadian cities, including Charlotte and Wilmington, North Carolina, and Ontario. The Australian Department of Defence has tested a wide-area camera in exercises over Adelaide (Au.) and Ottawa. MIT Lincoln Laboratory has flown its giant surveillance systems over downtown Boston. Harris has extensively surveilled the city of Rochester, and the Air Force has spent hours recording Ohio State University's campus in Columbus. Steve Suddarth's various flights over Albuquerque have been used by researchers at the University of Missouri to test a vehicle-tracking algorithm. In Pasadena, he flew a test flight over a crowd of tens of thousands of people at the 2016 Rose Bowl.

Perhaps most alarming of all, in September 2007, as part of a $4 million exercise backed by the CIA and NSA called Bluegrass, the people of Lubbock, Texas, were surveilled by two all-seeing-eye cameras, plus two radars that can detect thousands of moving targets on the ground simultaneously, as well as a soda-straw airplane, for an undisclosed number of days. The aim of the operation was to come up with ways of blending multiple surveillance feeds into a single view of the battlefield (a topic we'll get to later in the book). A US Army lab that is in possession of the footage describes it as a "large" data set. It is also a detailed data set. In one clip that I reviewed, several cars are tracked along Flint Avenue, a quiet residential street.

The CIA proposed conducting a follow-up to Bluegrass over an undisclosed US city in 2014, with the goal of figuring out how to better mesh traditional intelligence sources with social media feeds, but it appears the operation never materialized, or was never revealed.

Such flights continue to this day. In the summer of 2017, shortly before it was deployed to Syria, the Air Force 427th Special Operations Squadron's top-secret WAMI airplane spent more than 50 hours flying orbits over Seattle, in some cases loitering over a single neighborhood (Bellevue and Renton were favorites) for 4 or 5 hours at a time.

Even I have had a close call with a wide-area surveillance operation without knowing it. In 2014 I attended an art fair in Grand Rapids, Michigan. A week before my visit, the firm Consolidated Resource Imaging, which is based in Grand Rapids, flew a mock surveillance mission over the city to demonstrate how an all-seeing eye could be used for large public events.

When it comes to law enforcement, police are likewise free to use aerial video surveillance without a warrant or special permission. Under current privacy law, these operations are just as legal as an officer spotting unlawful activity while walking or driving through a neighborhood. Say, for example, an officer sees a grove of marijuana plants through the open window of a house. Since the cop is standing in public space — on a road or sidewalk — he or she obviously doesn't need "permission" to see the illicit plantation, or a warrant to photograph the scene.

The only caveat to police aerial surveillance activities is that they must employ a "publicly accessible technology," a term that has been defined, somewhat vaguely, in a small number of court cases. In two cases from the 1980s stemming from investigations in which police used daytime cameras aboard low-flying aircraft to spot marijuana plantations from the air, the Supreme Court had ruled that the law enforcement agencies had not violated the Fourth Amendment, since both helicopters and commercial daytime cameras are generally publicly available (if not exactly affordable).

In another case, *Kyollo v. United States*, which stemmed from an incident in Oregon in which police looking for indoor marijuana plantations had pointed an infrared camera at the plaintiff's garage from a public vantage point, the court ruled that the police's actions constituted a warrantless search, as an infrared camera most definitely did not constitute a publicly available technology.

The courts have yet to rule as to which side of the "publicly accessible" divide a 200-megapixel camera capable of watching a whole city at once would fall, but so far no law enforcement agency using WAMI has faced a legal challenge.

In case such a challenge ever were to arise in Baltimore, McNutt had compiled a memo outlining the legal basis for wide-area aerial surveillance in the United States. In effect, Persistent Surveillance System's operation in Baltimore was legally equivalent to a private citizen ("a guy with a camera," as McNutt put it) who submits photographic evidence of crimes to the police. "It's pretty much a slam-dunk case," he said, handing me a copy of the memo. Since the PSS cameras were not peering into people's homes, he explained, and since anything visible through the camera is visible from the sky, a public space, the company was, at least for the time being, in the clear.

Look Up

The laxity of US privacy law when it comes to the sky has not gone unnoticed by US law enforcement agencies, and the last few years have seen an astounding surge in persistent aerial surveillance over domestic soil by way of advanced police helicopters, soda-straw surveillance planes that continuously orbit targets in the fashion of Air Force Predators, and small multicopter drones — all of which is setting the stage for a future in which everybody is watched, all the time, from above.

On July 23, 2014, a Reddit user with the handle "jeandubrulee" wrote on one of the site's chatboards that a number of mysterious single-prop civilian airplanes had been flying tight circles over McLean and Langley, Virginia; one of the aircrafts' tail numbers was registered to a company called NG Research, but Google searches on the firm turned up little information.

Fellow Redditors following the discussion soon discovered the tail numbers of at least five other aircraft that appeared to be involved in the operation. One of the airplanes was sending out a transponder code — which air traffic controllers use to identify aircraft — that the FBI had been known to use for surveillance operations. Another aircraft, tail number N859JA, had spent several days flying a similar pattern over Quincy, Massachusetts, the previous year. No official explanation had been provided about the scope or intent of the surveillance in either case.

The following year, in May 2015, at the height of the Freddie Gray protests in Baltimore, a man named Benjamin Shayne observed a small civilian airplane circling over the city for hours on end. "Anyone know who has been flying the light plane in circles above the city for the last few nights?" he asked on Twitter. Seven minutes later, a Twitter user named Pete Cimbolic replied that he had found the airplane on a flight-tracking website. It, too, was registered to NG Research.

Cimbolic, a former employee of the American Civil Liberties Union, soon discovered that a second aircraft, a small corporate jet, had been tracing another orbit pattern over western portions of the city. He forwarded his discovery to the ACLU, which filed a Freedom of Information Act request with the FBI, DEA, and the Marshals Service.

Evidence logs released in response to the FOIA request showed that the FBI was behind the operations. It had begun aerial surveillance flights in Baltimore on April 28, three days after the first violent protests broke out, and had flown 36 hours of surveillance over the course of the following weeks. It had used both daylight and infrared soda-straw cameras, as well as electronic surveillance systems, the details of which it refused to disclose. NG Research turned out to be an FBI front company.

In response to the FOIA request, the FBI also provided an excerpt of its *Domestic Investigations and Operations Guide*, in which it asserts, like McNutt and others, that Fourth Amendment protections do not extend to the surveillance of citizens from public airspace.

The Freddie Gray surveillance operation was not in any way extraordinary. A former FBI agent who ran a special-operations group told me that aerial spycraft had long been part of the FBI's modus operandi. In testimony before the House Committee on the Judiciary that fall, FBI director James Comey explained that when investigators are unable to follow a target in a car or by foot, they will often resort to aerial surveillance. Asked about the extent of these airborne operations, Comey assured the committee that the bureau only operates a "small number of airplanes."

Comey's idea of "a small number of airplanes" is probably somewhat different from yours or mine. In April 2016 BuzzFeed reported that the FBI had used more than 100 aircraft to conduct 1,950 surveillance flights in a single four-month period in the fall and winter of the previous year. Most of these flights had been operated by a series of front companies like NG Research, all of which had similarly generic names — NBR Aviation, PXW Services, and so on.

The FBI explained that it used its aircraft to chase high-priority individuals such as "terrorists, spies, and serious criminals." And, to be sure, the planes' flight paths did closely resemble the persistent-stare orbits that Predator pilots employed on insurgent leaders in the Middle East. (FBI spokesperson Christopher Allen told BuzzFeed that the planes may also spend extended periods of time circling over a city simply waiting for a suspect to come out of a building.)

And yet the number of flights dropped by over 70 percent on weekends, which seemed to suggest that many of the targets weren't necessarily the type of high-value suspects that require 24/7 surveillance. On Thanksgiving, the flights stopped almost entirely. Many of the operations took place over Muslim-majority neighborhoods, such as San Francisco's Little Kabul and an area in Minneapolis known as Little Mogadishu.

The FBI is not alone among federal law enforcement agencies in its expanding use of aerial surveillance. The Department of Homeland Security operates a large fleet of surveillance planes, many of which appear to have been used for missions in support of other state and local agencies.

Customs and Border Protection, an agency within the Department of Homeland Security, is likely to be the first US federal agency to acquire WAMI for routine operations. The agency, which maintains a small fleet of unarmed soda-straw Reapers for border and maritime surveillance, has held a strong interest in wide-area surveillance ever since the Lawrence Livermore team demonstrated one of its early cameras on the southern border. Representatives from

DHS even attended some of the CIA's regular wide-area-surveillance conferences.

In 2012, after a DHS official met John Marion at one of the conferences, the agency hosted a demonstration of a Logos surveillance blimp at Nogales. On the first night of the operation, the camera led to the capture of 35 smugglers and migrants. By the end of the week, the authorities had apprehended more than 80 people using intelligence gathered from the camera. More significantly, over the course of the test period, the flow of people moving through the area had tapered off to a trickle. The organizers concluded that groups on the other side of the border had begun diverting their routes to avoid the blimp.

Two of CBP's Reapers are already equipped with a Vehicle and Dismount Exploitation Radar, a wide-area ground-surveillance system that was developed, like WAMI, as part of the Pentagon's counter-IED project in order to track insurgent forces roving in vehicles and on foot throughout the countryside. According to CBP officials, the radars, which also allow operators to peer deep into Mexican territory without leaving US airspace, have spotted more than 30,000 people crossing into the United States illegally, and have led to the seizure of over 12,000 pounds of cocaine. CBP has already invested in a number of efforts to process the data from a wide-area surveillance system, a strong indicator that it plans to buy into WAMI sooner or later. (If it hasn't already: one Department of Homeland Security document from 2013 refers to a single drone, also based in Sierra Vista, Arizona, that was equipped with a "Wide Area Surveillance System." No other CBP document appears to refer to the system, which is described as being capable of scanning an area six kilometers wide.)

Aside from DHS and the FBI, the Marshals Service maintains a secretive task force known as the Air Surveillance Branch, which provides "vital intelligence during the investigation and arrest of some of the country's most dangerous fugitives," according to the agency's website.

Even state and local police forces are increasingly watching their communities from the sky. In 2017 reporters from BuzzFeed ran US air traffic data through an algorithm that detects flight patterns commonly associated with surveillance activities, like repeated orbits over a single location, and discovered dozens of new unreported programs. The team found persistent surveillance programs over Los Angeles and Orange, California, and Phoenix, Arizona, among other cities. An investigation published by the *Texas Observer* in 2018 found that between January 2015 and July 2017 a single agency, the Texas Department of Public Safety, conducted hundreds of flights over the state's urban areas. These included a number of missions that appeared to have ventured over, or at the very least peered beyond, the Mexican border, as well as 38 sorties over Austin and 274 flights over the town of Roma, population: 10,265.

None of these operations have been widely publicized by the officials who approved them, and some have been actively concealed. The Palm Beach County Sheriff's Office in Florida has been found to have run a least one secret surveillance plane through a shell company called Five Point Aerial Survey for several years.

With the advent of commercial drones, even the smallest police departments can now put an eye in the sky. As of mid-2018 there were more than 600 state and local law enforcement agencies operating drones in US airspace. These small unmanned aircraft have only a fraction of the endurance and imaging power of manned aircraft, but they can serve as a powerful surveillance tool all the same. And soon, even an inexpensive drone might be all it takes for a law enforcement department to cast a wide surveillance net over an entire metropolitan area, a possibility I'll examine in the pages ahead.

Before long, US skies will also be home to larger drones capable of conducting military-grade surveillance. Predators and Reapers have already proven themselves to be unsurprisingly useful in domestic operations. In the summer of 2011 a Department of Homeland Security Reaper helped resolve a lengthy armed standoff involving four cattle thieves at a remote ranch in Lakota, North Dakota. Using surveillance footage from the drone, a SWAT team was able to descend on the property and arrest the men without incident. The California National Guard has, in turn, deployed Predators and Reapers to aid in the fighting of wildfires, and once used one of the aircraft to help find a man who went missing in Eldorado National Forest.

These operations took place in heavily controlled airspace where the risk of a collision between the drones and manned aircraft was minimal. But the Federal Aviation Administration, NASA, the Pentagon, and a cohort of large military defense contractors and industry groups have been working to develop collision-avoidance technologies, air-traffic-management systems, and policies that would enable large drones to have, as one Pentagon planning document put it, "routine" access to the same airspace that you or I use when we fly commercial.

The defense contractor General Atomics Aeronautical Systems has already optimized a number of its Reaper drones for use in civilian airspace, and company officials say they plan to have the Reaper fully certified for unrestricted flights in domestic skies by 2025. In 2018 a variant of the Reaper called the SkyGuardian crossed the Atlantic through entirely public airspace. The SkyGuardian has enough carrying capacity for a Gorgon Stare–sized camera, and it can stay airborne for more than 30 hours. Just imagine what your local precinct could do with one of those.

A Shooting, Plainly Seen

On my final day in Baltimore, McNutt invited me to a briefing that the Community Support Program had prepared for a team of three BPD detectives investigating a recent murder. At approximately four thirty in the afternoon on July 10, police discovered a 31-year-old man, Robert McIntosh, who'd been shot several times on a quiet block in a neighborhood known as Madison Park. McIntosh died on the way to the hospital, and no witnesses to the crime had come forward. But the camera had been watching.

In the footage from McNutt's airplane, McIntosh can be seen standing with a group of five men on the sidewalk. Suddenly, McIntosh is lying on the ground and the group disperses. One of the men, whom the analysts believe was the shooter, runs down a side street, while three others walk away calmly to the southeast. A fifth individual gets into a silver sedan, circles around the block, and picks up the alleged shooter in a nearby alley.

Tracking the sedan backwards in time, the analysts found that it had begun its misadventure outside a public school 45 minutes prior to the shooting. The car had taken a meandering route through the

A satellite image showing the coverage of the Community Support Program's surveillance orbits over Baltimore. The BPD would direct Persistent Surveillance Systems to either the eastern orbit or western orbit depending on where more crimes were expected on any given day. Each orbit captured 32 square miles of the city. *Google Earth*

area, at one point pulling into a dead-end street to rendezvous with a second vehicle, a black SUV. The sedan had proceeded to a parking lot, where McIntosh's shooter was waiting. The driver and the shooter then joined McIntosh and the other men on a sidewalk one block away.

After the shooting, the alleged shooter and the driver took a long drive through the city, doubling back on themselves several times — behavior, McNutt noted, that the analysts often observed in cars escaping crime scenes. Eventually, they joined the rush hour traffic inching along West Pratt Street, a major thoroughfare that cuts through downtown. A couple of minutes later, the sedan drove along a street that I recognized. It was half a block from the Community Support Program analysis center where the briefing was taking place.

The analysts had unfettered access to Baltimore's network of CCTV cameras, which they used to supplement their aerial imagery. If they lost a visual track on a vehicle from the air, they would turn to the footage from nearby ground-level cameras to find it again. From the CCTV images, one of the detectives was able to determine that the car was a late-model Chevy Cruz. The CSP team had also synced up their footage with CCTV feeds from around Madison Park showing close-ups of the remaining suspects who had been with McIntosh at the time of the shooting.

When the analysts came to a slide showing the car sitting in traffic on West Pratt, another detective noted that it was beside one of the city's public buses, which had recently been fitted with external CCTV cameras. He suggested that the analysts check with the city's transport authority to see if the bus had picked up a good image of the car's license plate number.[*]

All told, PSS had provided the investigators with the make, model, and year of a suspect's car, several addresses, and a rough narrative of the shooting, all of which would have taken the detectives much longer to piece together by any other means.

"This must be the best presentation I've seen in twenty-five years," said one of the detectives. "This is huge." He asked why every city in America wasn't using one of these systems. "A lack of political will," said McNutt, who then asked that the detectives relay their praise to their supervisors.

The detectives stood up to leave. It would later surface that other BPD investigators had declined to act on evidence collected by McNutt's team because the state's attorney had not been made aware of the existence of the program and any aerial imagery presented in court might therefore be deemed inadmissible. But the detectives on the McIntosh case seemed to be eager to use the leads presented in the briefing. "It's like that movie," said the one who had identified the car, shaking his head. He paused, trying to remember the name. Then, his face lit up: *Enemy of the State!"*

Hidden in Plain Sight

After the detectives left, McNutt pulled up a new video feed on the projector. The plane was airborne that day, and now the imagery

---

[*] License plates aren't visible from bird's-eye-view cameras. So here's a prediction: by 2030, cars will be required to have a third license plate — on their roofs.

we looked at was live. He zoomed into the troubled neighborhoods where he and his analysts focused most of their attention. I was startled by how these areas looked no different from the affluent parts of the city that CSP largely ignored. It is a funny thing about looking down on the world from above: nothing looks particularly suspicious — but by the same token, *everything looks somewhat suspicious.* Unprompted, McNutt scrolled over to the location of the CSP office, and for a couple of minutes we watched its empty rooftop from 10,000 feet in the sky.

Practically nobody else in Baltimore knew that a spy plane had been watching their every move for the better part of six months. Since the project wasn't funded from Baltimore's own budget, the Community Support Program had not been subject to the city's regular approval process, and BPD had not disclosed the existence of the program to the city government, the Maryland state legislature, or the public defender. BPD would later claim that because the flights were merely an extension of its CCTV surveillance system, which had already been green-lit by the city government years earlier, the airborne-surveillance project did not have to be run by any city official beyond the police commissioner. In this way, it was somewhat reminiscent of the rogue NSA cell in *Enemy of the State.* Even the mayor was kept in the dark.

It is not unusual for cities that test new and controversial surveillance technologies to keep these efforts from public view. In 2016 it was revealed that the City of Philadelphia was running a number of SUVs equipped with automated license-plate readers; the vehicles were disguised, poorly, as Google Maps cars. An increasingly popular way of obscuring secret programs is a technique known as "parallel construction," whereby investigators who use an undisclosed surveillance tool to uncover a crime will then "rediscover" the evidence using a known surveillance tool they are free to discuss openly in court (Human Rights Watch and other advocacy groups claim that this practice violates principles of fair trial and other basic human rights). In some cases, police departments have even opted to drop charges against suspects rather than appear before a judge who might require them to divulge the existence of a secret program like CSP.

And, just like CSP, most of the early efforts to use WAMI domestically have been attended by spy-like guardedness. Los Angeles's weeklong wide-area-surveillance pilot program in 2012 wasn't revealed to the public until 2014, by the Center for Investigative Reporting. When the story broke, the police sergeant who oversaw the program explained that if the police had disclosed the existence of the tests at the time, public pushback could have grounded the initiative before it even had a chance to prove itself. In 2018 Consolidated Resource Imaging was awarded a contract by a state agency to maintain a wide-area system on call for major events, security operations, and emergencies, but the company has not disclosed which state awarded the contract, and no state immediately came forward to announce itself as the customer. Even many relatively benign WAMI operations are needlessly obscured from public view.

McNutt told me that "the politicians" — meaning the BPD leadership — didn't want to reveal the existence of the Community Support Program until they had "six or seven" solved murders to show for it. He said that, for his part, he opposed BPD's decision to keep his work in the shadows. He wanted violent offenders in the city to know that the plane was overhead, and that it took his team only two hours to catch up to the men on the motorbike who had assaulted a police officer earlier in the month. Knowing about the 192-million-pixel camera staring down on the city, people would think twice before committing an offense. He was confident, too, that "90 percent" of Baltimore's residents would approve of CSP, and the other 10 percent could be easily convinced of its merits.

Even so, before I left the CSP operations center, McNutt instructed me, once again, not to speak to anyone about the program. Out on the street, I walked for a few blocks along the route that the silver sedan had taken the day of McIntosh's murder. It was midday, and there wasn't a cloud in the sky. I couldn't see the PSS plane, though I knew it could see me, and I felt its presence intensely. Much more discomfiting, however, was the sight of all the other people on the street, going about their business, unaware that they, too, were being watched.

5

Pixels into Ploughshares

WALKING THROUGH THE streets of Baltimore under an all-seeing eye, it's hard to imagine that this formidable technology, forged as it was for military use, might someday be regarded with the same unqualified public approbation as fire trucks or medevac helicopters. From the moment it was conceived, however, long before a single drop of blood had been shed on the battlefield under its shadow, its creators have insisted that it should, could, and would be a force for good in the world. When the Constant Hawk and Angel Fire engineers sat on the beach in Florida during their joint development work in 2005, they agreed, according to the contractor Nathan Crawford, that the technology "has to serve a positive purpose."

The technology could search for survivors following natural disasters, fight forest fires, improve people's morning commute, and even reduce insurance premiums. In fact, it has already proven itself to be just as adept in all of these roles as it was in unraveling insurgent networks in Baghdad or chasing murder suspects in Baltimore. There are, in short, many good reasons for those who dread aerial surveillance to temper their fear and mistrust.

Some of the earliest instances of wide-area surveillance over the United States were, in fact, wholly benign. In the summer of 2008, mere months after completing work on its first cameras, Persistent Surveillance Systems provided an eye in the sky for a Fourth of July celebration at Fort Leonard Wood, several college football games in Ohio, and Sarah Palin's vice-presidential-nomination acceptance speech.

In 2010 Consolidated Resource Imaging, the Constant Hawk contractor, flew a wide-area surveillance system over the Gulf of Mexico following the *Deepwater Horizon* oil spill. The camera tracked dozens of response ships and containment booms simultaneously. Following the operation, the Coast Guard reportedly estimated that CRI's lone airplane had been able to do the job of 10 helicopters equipped with narrow-field (soda straw) cameras.

It isn't difficult to imagine what this could mean for disaster operations. Over the course of the response to Hurricane Katrina, rescue helicopters clocked thousands of flight hours searching for the estimated 60,000 stranded survivors who waited patiently for deliverance. A single wide-area camera could have searched an entire neighborhood in one sweep. In one exercise off the Scottish coast in 2016, a solitary albatross-sized catapult-launched surveillance drone equipped with a wide-area camera was able to peruse every square inch of an area the size of Wales in just 55 hours.

While it searches for survivors, the wide-area camera's data could also serve other purposes. In 2013 Sierra Nevada Corporation flew a Gorgon Stare prototype over an area in Colorado that had recently endured some of the heaviest flooding in the state's history. According to Mike Meermans, the Sierra Nevada executive, the Federal Emergency Management Agency used the imagery to map out some of the affected areas. Persistent Surveillance Systems has carried out similar operations in response to flooding in Iowa in 2008 and in New Jersey and Pennsylvania following Hurricane Sandy in 2012.

In 2018 the Indiana National Guard announced in a short online post that a Gorgon Stare system had been employed as part of the relief effort in North Carolina following Hurricane Florence. This was the first publicly acknowledged instance of Gorgon Stare being actively used for an operation over American soil. The scant information that the Indiana National Guard provided suggests that the system had performed admirably, demonstrating a side to the technology that would have been alien to those who had used it in the context of war. According to the commander of the intelligence squadron that had used the Gorgon Stare imagery, it had been used for checking on the condition of critical infrastructure, identifying blocked roads, and directing rescuers to survivors in peril.

Anticipating similar needs on its own shores, the Australian Department of Defence optimized its first wide-area-surveillance airplane, a plain white civilian turboprop mounted with a variant of the US Air Force's Angel Fire, for disaster relief efforts — in addition, of course, to counterterrorism. (Unlike their US counterparts, Australian officials seem to have been less interested in names that might strike fear into the hearts of their enemies. Their WAMI program is called Wide Area Surveillance Activity Based Intelligence; for short, that's WASABI.)

One type of natural disaster for which Australia might find WASABI's continuous stream of data to be particularly useful is wildfires. The US Forest Service maintains an active wide-area surveillance program that has already demonstrated the technology's utility for firefighting at a time when fires are becoming more common and more severe.

Zachary Holder, a Forest Service official who manages the agency's WAMI program, often couches firefighting in the vocabulary of war. In his world, the aircraft that drop fire suppressant are like attack planes — you want to make sure they hit the right targets. The smokejumpers are like troops on the ground, in constant danger of becoming pinned down. The fire itself is a living, breathing adversary that you will only truly comprehend, he says, if you figure out its "pattern of life." Using the wide-area camera, Holder's unit can manage all of these dimensions of the fight simultaneously. It can ensure suppressant drops are hitting their marks, keep close tabs on smokejumpers, and anticipate the fire's behavior as though it were a network of insurgents. At the end of an operation, the footage can be used for after-action evaluations to hone units' firefighting strategies. Holder calls the technology a "game changer."

In 2017 the Forest Service awarded CRI a $4 million contract to lease a new infrared camera 10 times larger than the one it had used for testing. Holder estimates that the service would need only five airplanes equipped with this sensor to put a wide-area eye in the sky over all of the US wildfires that might need one in a typical season. If Holder has his way, thousands of acres of land, not to mention more than a few firefighters, could be spared from a fiery demise.

Taking to the Road

Other uses for WAMI are somewhat less heroic, though potentially no less gainful for those both on the ground and in the sky. While conducting overwatch for the Coca-Cola 600 NASCAR race in North Carolina in 2008, Ross McNutt and his team weren't nearly as busy as they might have liked to be. The highlight of the operation was when a call buzzed in over the radio asking if anybody had seen a man in a golf cart who had been involved in an argument with a police officer. Zooming in on the footage, the analysts followed the offender back to his trailer, where security guards found him and asked him to leave.

Looking for something else to do as the day wore on, the team started tracking the hundreds of vehicles that appeared to be roving aimlessly around the venue's vast parking lots in search of open bays. One particularly ill-starred driver spent nearly two hours searching for a space.

In contrast to the drivers' limited vantage point, it was easy to spot the Speedway's empty spots from the sky. McNutt realized that this information could be used to better direct incoming traffic to the emptiest lots. He calculated that he could spare thousands of cars more than an hour of searching. He believes that if each of these visitors had bought a soda in the additional hour they might have otherwise spent at the venue, the track would have made an extra $250,000 in profit. His fee for surveilling the three-day event was less than one-tenth of that.

McNutt's math might be somewhat optimistic, but the idea that WAMI, by providing a bird's-eye view of the traffic below, could help untangle a city's snarled roadways is credible. Traditionally, the people who study and manage traffic rely on complex theoretical models to describe driver behavior. These models are crucial for answering all kinds of vexing questions, like how to design a parkway that accounts for the frequency with which drivers will change lanes, or how to time the traffic lights at intersections to best accommodate the volume of traffic coming from each direction. These models are often far from perfect, in part because they are usually based on small sample studies rather than full-scale surveys—much like a political poll that predicts the voting preferences of 120 million people based on a study of 500 respondents who still use landline telephones.

To address the problem, two researchers at Virginia Tech, Kathleen Hancock and Rauful Islam, have been using a reel of wide-area footage of downtown Hamilton, Ontario, encompassing thousands of vehicles to better calibrate a number of widely used traffic models. So far, their results have been encouraging. Compared to sample studies, the researchers say, wide-area footage is like a census. With a precise narrative schematic of how each of the thousands of vehicles moving through a city chooses its route, a traffic authority could identify trouble spots, map out exactly how behaviors at intersections affect traffic elsewhere, and much more accurately anticipate the impact of disruptions like road closures or new construction projects.

There's more. From above, traffic violations are very easy to spot. In one three-minute clip of footage from a midsize city that I reviewed, I counted numerous violations that would have otherwise gone undetected, such as drivers joining the wrong lane when turning at an intersection. The footage from Hamilton, Ontario, is said to abound with cars ignoring traffic lights and stop signs. Over the course of the surveillance program in Baltimore, Persistent Surveillance Systems analyzed 42 traffic accidents and was able to identify who was at fault in all but 7.

The implications for municipal revenue from traffic tickets, not to mention the rapid settlement of disputes between drivers following an accident, is clear—as is the potential upside for criminal justice. In 10 cases, analysts were able to identify drivers who had fled the scene of a hit-and-run.

Something for Everybody

To be clear, WAMI is not cheap, and that puts the technology out of reach for many. The operation in Baltimore cost $120,000 a month, and that was a discounted rate. Ross McNutt's full monthly fee, which gets the client about 100 hours of daytime surveillance, is $200,000.

The traffic-surveillance flights over Hamilton reportedly cost about $5,000 an hour. The US Marshals Service, the Secret Service, and the FBI, among other agencies, have all taken an interest in Vigilant Stare, Sierra Nevada Corporation's civilian version of Gorgon Stare, but in each case the camera's cost was a deal breaker. (Vigilant Stare's price is undisclosed, though it is probably within range of Gorgon Stare, which costs about $20 million per unit, not including the airplane to carry it.)

But if a city was to use a wide-area surveillance aircraft as a shared tool, accessible to any authorized party who needed an eye in the sky — the police, the fire department, the traffic authority, and so on — and if the cost was likewise shared among these agencies, that price tag seems much less daunting. After all, that was part of the idea behind Gorgon Stare — instead of serving a single unit on the ground, a solitary Reaper could serve dozens of operations concurrently. While a firefighter watches over a multiple-alarm blaze, a city planner might be examining traffic congestion on the other side of town, while the police keep eyes on a neighborhood parade. At the end of the day, everyone pays and everyone benefits.

If, even then, no individual city government were willing to pay for a perpetual eye in the sky, the technology could be owned and operated by state and federal organizations, which would dispatch the systems to areas where they are needed, the way the Department of Homeland Security dispatches a fleet of helicopters to major security events such as political party conventions. One week a Vigilant Stare might be assisting a disaster-response effort in Florida; the next, it's in DC for the presidential inauguration.

This is how Zachary Holder at the Forest Service envisions using his future fleet of WAMI planes, and it appears to be the model for the mysterious state agency that contracted Consolidated Resource Imaging to keep an airplane on call at all times. A number of the engineers and salespeople involved in the industry believe this will be the winning bet.

And that is assuming that the use of the technology would be limited to public sector applications, which it wouldn't. There is no shortage of potential uses for an all-seeing eye beyond the government, and no reason, at least from a practical or legal standpoint, that the private sector be kept out of the picture. Major US insurers have already contracted a number of companies to scan areas affected by natural disasters in order to record property damage on their claimants' homes. In one such operation, in 2016, the firm Acorn surveilled large swaths of northern Florida following Hurricane Matthew for an unnamed underwriter.

Working with CRI, the insurance industry has also explored the idea of using the technology for sniffing out common fraud schemes, such as sham medical clinics that file insurance claims on behalf of car-accident victims. From the air, it is fairly easy to distinguish the sham clinics, which dispatch employees to accident scenes in the hopes of finding victims to dupe, from legitimate medical institutions, which wait for victims to come to them. Since these cases often involve organized crime groups, neither CRI nor the insurers are willing to go into detail about the pilot programs they have already run in several "hot spots." No one wants to wake up to a horse's head in their bed.

Wide-area surveillance also has the potential to be a powerful business intelligence tool. In June 2016, for example, the citizens of Wilmington, North Carolina, became the unwitting subjects of a three-week wide-area-surveillance test by Acorn, a contractor with deep military ties hoping to demonstrate how the technology could be used to track traffic patterns at retail stores. The company believes that it could sell this information to retailers hoping to better understand where their customers live and work. As of this writing, the company has not replicated the operation in Wilmington, which it says was a test.

Meanwhile, large energy and utility firms want CRI and other wide-area surveillance firms to fly daily flights along certain underground pipelines searching for construction machinery that could be digging dangerously close to the buried infrastructure. A laboratory at the University of Dayton is developing computer-vision algorithms that can do this automatically.

An airborne-surveillance company such as Acorn could provide the feed from a single camera to a range of customers simultaneously: an insurer might settle a who-hit-whom dispute at one intersection while a real estate developer tracked foot traffic around a commercial lot. A bank could analyze that same foot-traffic data to decide whether or not to give said developer a loan for the property.

And then, of course, there are all those companies and people who will figure out how to use the all-seeing eye in ways nobody has yet imagined. In various conversations, WAMI was described to me as being like Google Maps or the iPhone — a technology with far more uses than its creators could have possibly imagined themselves. Some of the eventual uses that will be conceived could be dangerous and harmful. But the industry is convinced that many others would no doubt be of great benefit to society. Some might be very profitable. Others, they all say, will save lives. And if they do, those men and women who spent so many years toiling in obscurity to bring WAMI to light may yet have their own day in the sun.

6

What's Next for Wide-Area Surveillance

AS WE BEGIN to explore the many possibilities, both peaceful and not so peaceful, of the all-seeing eye, we must keep in mind that we have so far witnessed only the beginning of what will very likely be a long technological evolution that unfolds in incredible and even frightening ways.

When Gorgon Stare was completed, Michael Meermans, the Sierra Nevada executive who helped launch the program, felt proud that his team had, as he put it to me, "fundamentally pushed, not broken, but pushed at the bounds of physics." But, he said, "At some point, is that the end?

"No. When it comes to the world of actually collecting information and creating knowledge, you can never stop." One must always strive to watch more targets, more persistently, more clearly, and at greater fidelity. Surveillance is a long road.

Here's what we know is coming next: The cameras themselves are becoming more powerful. Their size and cost is shrinking while the number of countries and organizations that possess them is growing.

Most important of all, though, the technology is becoming intelligent.

Bigger, Better, Smarter, Smaller

Early proponents of the all-seeing eye often found themselves arguing with officials who were certain that the one million pixels in the Predator's camera was more than enough for the Pentagon's counterterrorism surveillance needs. *Why would anybody need 100 million pixels?* they would ask. A commander in 2018 has almost certainly said the same of Gorgon Stare. *Two thousand million pixels? Surely that's more than enough!* But, the long road . . .

Indeed, the pixel-count record set by ARGUS has already fallen. Under a separate DARPA program called AWARE (which stands for Advanced Wide FOV Architectures for Image Reconstruction and Exploitation), researchers at Duke University and a number of other labs built a series of progressively more powerful cameras, culminating with AWARE 40, which has 40 gigapixels and 27 times the acuity of a human eye.

There are many factors that go into the making of a more powerful camera. One of the most important is the size of each pixel sensor on the photographic chip. With smaller pixels, you can collect higher-resolution images without having to make the chip itself — and the camera, in turn — any bigger.

To give you a sense of what that means for future eyes in the sky, here's a little math. The individual pixels on the cell phone chips used in ARGUS and Gorgon Stare are 2.2 microns wide; with five million pixels, the chips are about five millimeters wide. If the pixels are reduced to 1.22 µm, which is the size of the pixels in an iPhone 8 camera, you could, theoretically, build a Gorgon Stare that's roughly the same size as the original but five times more powerful.* The computers that process all of these pixels are also becoming cheaper and smaller. In 2012 it took one of Livermore's famous supercomputers to run the lab's processing software for Gorgon Stare. Now, the program could be handled by a single Apple laptop.

As a result of these advances, WAMI systems have been shrinking. The first wide-area-surveillance cameras alone, without any supporting components like the computers and gimbals, weighed between 100 and 200 pounds. The first version of Constant Hawk was so large that it barely fit in an airplane designed to carry 39 passen-

---

* The falling cost of high-power digital-imaging technologies, along with the advent of stitching software, has given rise to a large online community of hobbyists with, as one site, GigaPan, puts it, "a passion for gigapixel photography." Many of the images uploaded to the site, which are created by stitching together hundreds of small high-resolution images to create a single gigantic frame, now exceed a hundred gigapixels. A few even have a thousand gigapixels. Welcome to the age of the terapixel — that is, a *trillion pixels.*

gers. Gorgon Stare was so heavy that the Reapers installed with it had to forgo their usual complement of precision missiles and bombs; the crew had to rely on other aircraft to strike targets identified in the footage.

The most recent WAMI cameras are, by comparison, tiny. The Marine Corps is preparing a WAMI system with the code name Cardcounter for a small "tactical" drone, probably the RQ-21 Black-jack. The camera will be capable of watching a 16-square-kilometer area — as much as the first Gorgon Stare — during the day or night, but it will do so from a drone that is more than 40 times lighter than the Reaper. Logos Technologies' Redkite, a 70-megapixel camera capable of surveilling 12 square kilometers, weighs just 23 pounds.

Eventually, though, attempts to further miniaturize pixels will hit a wall, as they will always need to be large enough to at least absorb individual photons. So researchers are turning to less direct ways of getting more out of their cameras. One option is to use a low-resolution camera that scans the entire surveilled area between each frame, creating a composite image, much like when you take a panoramic photograph with your smartphone. Agile Spotter, a 3.2-megapixel unit designed by Logos, takes 35 individual adjacent images each second, allowing the relatively humble sensor to generate a seamless 115-megapixel mosaic of footage covering a broad area.

Another technique, a trick known as "superresolution," which uses computer-vision algorithms to correct blurriness and graininess in image frames, can turn small indecipherable blobs in a low-resolution aerial video into crystal-sharp cars, buildings, and people. The effect is similar to the fictional "enhance" feature seen in action films and TV shows, where a character (usually of the nerd archetype) is magically able to sharpen grainy images with a few keystrokes. A real-life superresolution tool developed at the University of Dayton's Center of Excellence for Computer Vision and Wide Area Surveillance Research can turn 40 pixels, which looks like nothing more than a smudge on the screen, into 640 pixels: a nice, crisp image of a car.

The Dayton Vision Lab and other groups are also creating software that automatically sharpens shadows in airborne footage, which further enhances the clarity of the image. The results are striking even to the untrained human eye—targets become easier to track and enemies become easier to find.

Another, much more radical, approach is to essentially give each pixel a tiny brain. That's the idea behind something called the digital-pixel focal plane array, an infrared video chip in which every pixel is attached to a converter that digitizes the incoming light right on the spot. (Traditional cameras, by contrast, digitize the entire image on a single converter.) At the moment, DFPAs can only be used for infrared cameras, which have, according to the engineer Bill Ross, hit their own miniaturization wall at the 10-micron mark, but as converters get smaller, the technology will find its way onto daylight cameras, too—eventually, it may even become a feature on regular smartphones.

Since each pixel can more quickly digitize what it's seeing, you can put the camera on a fast-spinning mount that turns across the whole field of view between each frame, like the Agile Spotter but on a much larger scale. One DFPA camera, the infrared AirWISP, developed by the MIT Lincoln Laboratory with support from the CIA, spins a thousand times a minute, enabling it to monitor an area the size of Pittsburgh in a single frame—far more than equivalent infrared cameras of the same size and weight.

With roughly the same computing power as the systems that guided the first mission to the moon, each pixel can also correct motion that would otherwise blur the image, eliminate the image background so as to better capture fast movements—one test showed that it can even catch a bullet speeding through the air—and pick up extremely faint signals that would have been invisible to an ordinary pixel. All of which is useful if you're trying to watch thousands of elusive targets moving quickly across a wide area from thousands of feet up in the sky.

Going Global

In 2017, a few months after Logos revealed that it had successfully mounted its Redkite sensor on a Boeing Insitu Integrator drone and flown it from a snowy airfield in Oregon, Redkite was named "Best New Product" by *Aviation Week & Space Technology,* an award that

A Boeing Insitu Integrator catapult-launched drone equipped with Redkite, a miniaturized WAMI system, during testing in Boardman, Oregon, in 2017. Redkite is 40 times lighter than the early WAMI units, but just as powerful. *Insitu Communications*

is as close as anything to an Oscar in the aerospace industry. The industry has good reason to celebrate the Redkite. Unlike the large and expensive drones that were traditionally required to carry a WAMI, the Integrator is small and (relatively) cheap. Since it can be launched by catapult, it doesn't require a runway, so it can be operated from ships and remote areas.

More importantly, because the Integrator is relatively benign compared to something like the much larger Reaper, which is equipped with sensitive — and lethal — technologies, the United States is happy to sell it to a long roster of even distant allies. Canada, the Netherlands, Poland, and the United Arab Emirates already operate Integrators. Afghanistan, Cameroon, the Czech Republic, Iraq, Kenya, Lebanon, the Philippines, Pakistan, and Tunisia — all countries that may never be allowed to buy Reapers — have acquired fleets of ScanEagles, a drone similar to the Integrator manufactured by the same company. Previously, WAMI would have been off-limits to these militaries. Not anymore.

As the barrier to entry in wide-area surveillance continues to fall, international demand is rising. There is certainly no shortage of regimes that would welcome the opportunity to acquire the technology, not always to noble ends. The Philippines could use it to hunt members of the jihadist organization Abu Sayyaf. Myanmar could track Rohingya refugees. Turkey's government might want an all-seeing eye to monitor protesters in Istanbul or direct airstrikes against the outlawed militant group PKK in Kurdistan. Hungary might adopt it to watch over its borders, looking for migrants fleeing the violence in the Middle East. A "wide-area surveillance capability" is an increasingly standard requirement in military tenders for new spy planes.

The principal wide-area surveillance companies have long been positioning themselves to meet that demand beyond the US market. As of the spring of 2018, Logos was aggressively pushing to provide sensors for a fleet of surveillance blimps destined for Saudi Arabia and the UAE. Since at least 2015, Harris has been pitching the CorvusEye 1500 to countries and law enforcement agencies of the European Union. Though CorvusEye is subject to export restrictions that the US government imposes on sensitive military technologies, it could nevertheless be purchased by up to 47 foreign governments. (Meanwhile, PV Labs, a Canadian outfit, sells a 300-megapixel camera that the company notes is completely free from US export restrictions.) At least one European airborne-surveillance firm — the Dutch company Vigilance — maintains an aircraft equipped with a Corvus-Eye to "quickly respond to threats."

The Israeli defense firm Elbit Systems, a relative newcomer to the wide-area market, sells a unit called SkEye that it claims can cover an 80-square-kilometer area, unblinkingly, for hours on end. It is already reportedly in use by several undisclosed militaries. Shortly after a run of terror attacks in the UK in the spring of 2017, the company went on a marketing offensive, telling news outlets that its system was ideally suited to investigating such events. "Wide-area surveillance" has become such a buzzword in the field that it is even used in marketing copy for traditional soda-straw cameras.

Various governments are also taking steps to develop their own wide-area airborne-surveillance capabilities in-house. The UK Ministry of Defence has quietly begun investing in wide-area surveillance technology, and has already sponsored a number of (unpublicized) test flights tracking cars along major roadways like the M1 for a computer-vision-based program that autonomously develops pattern-of-life portraits from wide-area-surveillance data. The German Aero-space Center is developing a wide-area airborne-surveillance system called, perhaps not coincidentally, ARGOS, a 48-megapixel camera array that will be used for traffic studies and to monitor natural disasters and large events. The governments of Germany and the Netherlands have likewise sponsored their own research initiatives in imagery processing for wide-area footage. Singapore is working airborne wide-area surveillance into its plans to become the world's first major smart city.

Farther south, the Australian Department of Defence has its WASABI plane. Aside from exploring the technology's potential for disaster response, engineers are using the system to develop image-processing software, while combat units are using it to hone their techniques for urban warfare operations. In 2017 and 2018 it brought the aircraft to Contested Urban Environment, a yearly counterterrorism exercise held by the member states of the Five Eyes intelligence partnership: the United States, the UK, Canada, Australia, and New Zealand. (The Pentagon also appears to be sharing its work on Project Maven — see chapter 8 — with its Five Eyes counterparts.)

Meanwhile, as of this writing, General Atomics Aeronautical Systems, the American drone maker, is seeking to equip the fleet of Reapers it is building for the Australian DOD with a wide-area sensor made by Sentient Vision. Unlike Gorgon Stare, the Sentient Vision camera weighs only a few pounds, leaving plenty of space for the drones to carry missiles — the first known armed all-seeing eye.

Those who lack such cozy relationships with the United States will seek to replicate its progress by less legitimate means. In 2015 the Russian hacking group Fancy Bear, which is better known for its intrusions into the 2016 US presidential election, targeted dozens of current and former Pentagon officials and defense contractors involved in the development of airborne-surveillance technologies, including General Jim Poss, who was hacked after opening a fake Google security notification email while attending an airshow in Paris. Another target of Russian hacking attempts was Clarifai, a software startup that worked on Project Maven. None of the targeted organizations and individuals is willing to disclose what information may have been filched, as most of the stolen material is classified.

Russia's campaign mirrored ongoing Chinese efforts to steal secrets from a number of America's largest defense contractors, including General Atomics. Although there has been no evidence directly linking these efforts to WAMI technology, China is, in any case, in a strong position to become a global leader in the field, with or without the help of economic espionage.

China's surveillance-technology sector is already experiencing a growth spurt that is unprecedented in the history of the industry. Hangzhou Hikvision Digital Technology, a Chinese firm with government ties that develops artificially intelligent CCTV cameras, has already sold surveillance systems, including high-resolution wide-angle ground sensors, to customers in more than 100 countries, including the United States. Scientists from the National Laboratory of Pattern Recognition at the Institute of Automation of the Chinese Academy of Sciences, a government laboratory in Beijing, have been publishing extensively on autonomous analysis systems for wide-area-surveillance footage, a clear indicator that the country soon plans to generate its own wide-area surveillance in need of analysis, if it doesn't covertly do so already.

China would certainly have a healthy domestic demand for the all-seeing eye. In the last decade, the Communist Party has blanketed the country with more than 200 million CCTV cameras and a wild assortment of other surveillance systems. It would also be in a position to find a healthy market among foreign militaries and regimes that would be forbidden from purchasing something like Gorgon Stare. Already, thanks to its looser export rules, China is fast becoming the world's largest military drone exporter; Chinese firms have already sold large armed drones (which would be capable of carrying WAMI sensors) to Turkey, Pakistan, Nigeria, the United Arab Emirates, Saudi Arabia, Iraq, and Jordan, among others.

Indeed, Jordan only turned to China after the United States rejected its requests to buy small fleets of American Predator drones for fighting ISIS. The next item on Jordan's surveillance shopping list might just be a Chinese Gorgon Stare knockoff.

Going Local

Someday, even your neighbor might be able to buy a WAMI system, as the technology will soon be small enough to fit on regular commercial drones that can be bought online for a few thousand dollars and flown with little more than 10 minutes of training.

One especially ominous portent of this future is the DragonFly, a 90-megapixel 360-degree camera built by the MIT Lincoln Laboratory that's mounted on the kind of commercial multicopter drone commonly used by the film industry for aerial cinematography. The camera is powerful enough to spot vehicles at distances of up to three miles in any direction; the drone itself is attached to a portable ground station via a thin tether, so it can remain airborne indefinitely. (The DragonFly was tested alongside Australia's WASABI plane in the Contested Urban Environment exercise in Adelaide in November 2017.)

Steve Suddarth, the engineer who developed Angel Fire, has an even more ambitious system in the works. He has built a drone-mounted 30-megapixel camera that's 1/1,000th the size, cost, and weight of Constant Hawk, and is developing software that stitches the footage from multiple such drones spread evenly over an area, the same way ARGUS stitches footage from its array of 368 camera chips, to produce a single video. With 76 drones, he has estimated that one could cover the whole of Manhattan.

As the footage would originate from multiple vantage points, the software will produce a highly detailed three-dimensional, video-game-like view of the ground, rather than the less intuitive map-like imagery produced by a solitary high-altitude camera. And since the drones would fly under the clouds, they could operate even when it's completely overcast.

The Pentagon is already experimenting with a similar concept under a program called Perdix. A Perdix drone is about the size of a hamster and carries a camera with the same number of pixels as a single imaging chip inside ARGUS. But Perdix drones can be launched by the hundreds from the belly of a fighter plane, and they are programmed to fly in a dynamic, intelligent swarm. Such a swarm could effectively serve as a single gigantic camera capable of scanning enormous areas.

When I approached the Air Force about its use of swarms as a distributed wide-area-surveillance system, I was told — firmly, but politely — that it would provide no information on the matter beyond what has already been made public, which is pretty much nothing.

7

It Takes a Million Eyes

NO MATTER HOW fast, sharp, or light you make your tools of surveillance, they all still rely on one crucial, invariable component: the human.

For the first 150 years of airborne surveillance, from the blimps of the American Civil War to wide-area systems that grew out of the war on terror, it has always taken a human to turn the pixels, codes, and pulses generated by the tools of spycraft into knowledge. Though there are a wide variety of mechanical ways to spy from the air, only a human can determine whether the tall man wandering around Tarnak Farms is, in fact, Osama bin Laden and not a tall bearded civilian who merely resembles him.

This task has never been quick or easy. During the Cold War, it took thousands of photographic interpreters to make sense of the images captured by the US government's fleet of spy satellites. Video analysis is an even more labor-intensive process. A single soda-straw-view Reaper requires eight video-imagery analysts to constantly review its feed. These analysts must pay unwavering attention to their screens. According to Steven K. Rogers, a senior Air Force scientist, they are told that they must ask permission to turn their eyes away for any reason — even just to sneeze. As one defense researcher who studied the issue extensively put it to me, "If someone makes a mistake, someone gets killed."

The more pixels and pulses you generate, the more humans you need, and no single surveillance device ever created generates as much data per mission as wide-area motion imagery. In its first few months in Iraq, Constant Hawk filled so many high-capacity Hitachi DeskStar external hard drives that the shipping container used to

Nathan Crawford pictured at an undisclosed location in Iraq organizing Hi-tachi DeskStar hard drives for Constant Hawk imagery, which was shipped by courier to various secret analyst cells in Germany and the United States.
Consolidated Resource Imaging/Nathan Crawford

house them buckled under the weight of all that information. There were rumors that, back home, after the Army's covert acquisition of several thousand DeskStars from a wholesaler outside Chicago, the price of the hard drives went up by a couple of dollars nationwide.

Even if that is pure Army mythos, what is true is that the data repository generated by Constant Hawk was one of the densest the National Geospatial-Intelligence Agency had ever managed. The later, more powerful systems generated even more data. Gorgon Stare likely produced, in a single mission, several dozen terabytes of footage, enough to fill up the memory of more than a few hundred modern MacBook laptops.

Nathan Crawford, the contractor who deployed with the Constant Hawk mission to Iraq, was the one who personally loaded the Desk-Star hard drives into the shipping container that collapsed. When colleagues joked that they were building Big Brother incarnate, he would disagree: "It really takes a million people to watch a million people," he liked to say.

This is only a slight exaggeration. If you wanted to display every single pixel from a Gorgon Stare on an iPad, you would need 2,358 iPads. A highly caffeinated team of 20 analysts will be able to process only about one-tenth of what a single system collects. To be sure, when you record such a large area, not all of the footage is necessarily relevant to the mission. But it is inevitable that large volumes of significant information will wind up on the cutting room floor, unseen by human eyes.

The Pentagon has been aware of the problem since the beginning of our story. In 2008, shortly after the first WAMI systems were deployed, it even put the Jasons on the case; in their report, the group recommended that the Pentagon tackle the issue with a series of "Grand Challenges," a type of open competition reserved for the government's most complex technological goals (it was a DARPA Grand Challenge in the mid-2000s that catapulted the development of self-driving vehicles into the realm of the possible).*

Simply hiring more staff to watch all the footage is not an option. Researchers at RAND Corporation, which the Air Force contracted the year after the Jason study to figure out ways to help the service keep its head above the coming data flood, calculated that the service would need 117,000 analysts if it wanted to scrutinize every inch of Gorgon Stare video. It would, of course, be impossible to recruit so many people for any Air Force intelligence job, let alone an assignment that was described by one Predator pilot as like playing "the most boring video game in the history of the world."

The Pentagon turned to the commercial world for ideas. In 2011 the Air Force and the NGA hired an engineer at a major US broadcaster with experience wiring up live coverage of major sporting events. The engineer, who asked to remain anonymous, brought groups of intelligence officials on a series of tours of more than a dozen NASCAR races and football games, as well as the broadcaster's studio in New York City; the experience was eye-opening for many participants, and though it inspired some improvements to the Pentagon's video transmission and archiving techniques, it was clear to all involved that the challenge posed by WAMI was in a league of its own.

---

* One of the self-driving car Grand Challenge events, in Victorville, California, was watched over by, you guessed it, a WAMI airplane. According to John Marion, the system, operated by Logos Technologies, was used to help judges track all the competing vehicles simultaneously.

Concurrently, the Joint Forces Command paid Lockheed Martin and a group of subcontractors $29 million for a software program similar to the one ESPN uses to catalog its vast archive of game-day footage. The software gives analysts access to a large searchable archive of surveillance material, but it requires the footage to be meticulously tagged by hand, which, again, requires an enormous amount of human energy. (Have you ever wondered how, when you're watching a baseball game on TV and the pitcher throws for a double, the network is able to magically conjure a clip of a similar play from five games ago? It's because an analyst in Major League Baseball's office in Manhattan's Chelsea Market, using a similar computer program, is tagging every single play manually. From what I've been told, it's not a great job.)

Ultimately, in what was either a remarkable feat of lateral thinking or an act of sheer desperation, the RAND researchers began to study reality TV. They hoped that by visiting an actual reality TV set in Hollywood and watching how the crew handled large volumes of mostly very boring video footage from a network of multiple cameras, they might glean some tips to help the Air Force handle its own barrage of mostly very boring video footage from the wars in Iraq and Afghanistan.

RAND signed a nondisclosure agreement with the production company that accepted its request for a site visit, though a study later published by the research team detailing how the Air Force could manage aerial-surveillance video as though it were *Big Brother* (the show, that is) acknowledges two TV producers who worked on *Rock of Love: Charm School* (featuring Bret Michaels) and *Boob Jobs and Jesus.* This was a detail that the RAND team probably left out of the PowerPoint presentation they delivered around the Pentagon to promote the concept in 2012. Unsurprisingly, after an initial flurry of interest in the study, the Air Force decided to take a pass on the idea.

There is, of course, another solution — give your eyes a brain of their own.

After all, we are already surrounded by machines that have become frighteningly adept at mimicking what Darwin described as the many "inimitable contrivances" of the human vision system. Google Street View's algorithms are capable of collecting the street number of every property in France in less than an hour. Self-driving cars can — usually — tell the difference between a pedestrian and a plastic bag blowing in the wind. Even cheap consumer drones come with a computer-vision-based "follow-me" feature, equally effective for filming yourself skiing down a mountain and following your neighbor to the gym.

Surely if a computer can do all this, it could do at least some of the work of an intelligence analyst.

Using computers to watch video in the manner of trained human analysts would have profound consequences for airborne spycraft. You could greatly expand the volume of surveillance you collect without needing new personnel to process it. Computers could watch the equivalent of thousands of screens of video at once, like an army of unblinking, artificially intelligent desk soldiers. They wouldn't get bored, tired, or distracted. They wouldn't need to sneeze or take bathroom breaks. But is it really possible to remove the human component?

The Track Pack Goes to War

In 2015 two researchers from Michigan State and Pennsylvania State who analyzed thousands of pages of photo-interpretation manuals from 1922 to 1960 found that imagery analysis essentially comes down to two core activities. The first, identification, refers to the simple task of identifying what you see in the image: when you see a tree, you call it a tree; when you see a squirrel, you call it a squirrel. The second is recognizing the significance of what you've identified, a much more complicated task we'll get to in the next chapter.

Identification continues to be one of the fundamental tasks of aerial spycraft in the 21st century, often in the form of tracking roving targets as they move around the battlefield. In wide-area imagery, "tracks" — the path of a vehicle or person between different locations — are like the metadata that the NSA collects on telephone calls. Like metadata, a track doesn't, on first impression, appear to tell you all that much about your target. It doesn't reveal the type of car that's being followed, who's inside it, or what they're saying. But it does tell you where the car went, for how long, and how it got there. This information can reveal whether the target is associated with other known combatants, as well as the physical location of the nodes where he prepares for his attacks and hides in the aftermath, all of which is crucial for "attacking the network."

To follow a single vehicle, wide-area imagery analysts click on the target as it moves in each frame of the footage — one click per frame. When the footage is played back, the vehicle's path is overlaid on the imagery as a brightly colored line that looks something like a fluorescent worm. "Brute force tracking," as it's known in the spy community, is relatively easy. Many of the Gorgon Stare analysts who spend their shifts brute-force tracking insurgents in Afghanistan and Syria are fresh out of high school.

The task is mind-numbingly tedious. But the stakes are high. You

Airmen assigned to the Indiana Air National Guard's 181st Intelligence Surveillance Reconnaissance Group analyze surveillance footage from a Gorgon Stare mission over North Carolina as part of the response to Hurricane Florence. Imagery analysis is difficult, tedious, and labor-intensive, especially when it comes to wide-area surveillance, and the margin for error is minimal. Hence the coffee cups. *The appearance of US Department of Defense (DOD) visual information does not imply or constitute DOD endorsement. US Air National Guard photo by Staff Sgt. Lonnie Wiram*

have to be sure that the car at the end of the track is the same car you started following at the beginning, so you can't skip a single click. Otherwise, you might be putting an innocent civilian in the crosshairs of an operation while the real target slips away. To follow a vehicle that traveled for 30 minutes in a Gorgon Stare feed, you would need to click through 3,600 individual frames. A single frame of wide-area airborne surveillance might include thousands of moving vehicles, including dozens of targets of interest. As a result, the analysis of footage from a mission often takes much longer than the mission itself. Monitoring a single node on the off-chance that it might receive a significant visitor is an even more tiresome exercise.

If it was possible to track vehicles and nodes automatically, the work that took Gorgon Stare analysts days might require only a few seconds of computer processing. Virtual perimeters — "watchboxes" — could be set around nodes of particular interest to automatically alert users whenever a visitor is detected. Thousands of targets and nodes could be tracked at once. The analysts' time could be spent on more complex tasks, like deciphering the meaning of a single vehicle's journey from the site of an ambush to a residence on the other side of the neighborhood.

The teams at Livermore, MIT, Los Alamos, and DARPA all began working on automatic-analysis software almost as soon as they began building the cameras themselves.

Many of the scientists involved had backgrounds in automated image analysis. Bill Ross, the Lincoln Laboratory engineer, spent years developing object-detection and -tracking algorithms for classified airspace and missile defense. Steve Suddarth had worked in artificial intelligence and machine vision since the mid-1980s. On a Pentagon exchange program in France, he collaborated with Yann LeCun and Léon Bottou, pioneering computer-vision experts who both eventually took positions as senior researchers at Facebook. Later, at the Ballistic Missile Defense Organization, Suddarth participated in an effort to build a sugar-cube-sized computer that identified incoming cruise missiles using sophisticated tracking algorithms. Guna Seetharaman, an engineer on the Angel Fire project, based some of the system's software on algorithms that he had originally developed to help guide early self-driving-car prototypes.

Some of the most groundbreaking early work was carried out by Sheila Vaidya, the Livermore engineer who used graphics cards from Xboxes to build powerful imagery-processing computers. For half a decade starting in the mid-2000s, Vaidya ran a program, sponsored by the Air Force and NGA, called Persistics, a neologism combining "persistence" and "integrated circuits" ("ICs," for short), which she founded with the goal of using computers to extract as much information as possible from the imagery being generated by her colleagues on the wide-area-surveillance team.

Exploiting the same qualities that enable graphics chips to generate huge virtual worlds in video games, Vaidya's team built software that could establish the precise location of physical features that are immobile, like buildings, trees, and the ground itself, and then pin the imagery to these 3-D models, creating a stabilized view of the environment.

Employing a similar method several years later, a camera built by Logos generated such stable imagery that a moving clock had to be added to the footage because, absent moving targets in the imagery, the screen appeared to be frozen.*

Once the imagery was stabilized, the Persistics computer would simply isolate and track the moving vehicles, which stood out sharply against the unmoving background. With this tracking ability, according to an article in *Science & Technology Review*, Livermore's in-house journal, an analyst would simply have to feed the system instructions, like "Give me the frames that recorded this vehicle from one to two o'clock this afternoon," or "Show me all the vehicles that stop at this location today." If a vehicle passed through a watchbox, the software would automatically begin tracking it both forward and backward in time (the engineers working on Sonoma had previously built a similar tool).

That wasn't all. The software could increase the video's resolution so that analysts could get a better lock on suspicious vehicles.

---

* There was a second motive for the Persistics stabilization: since the background didn't move, the computer would send down only footage of the moving pixels, which reduced the amount of bandwidth required to get live footage from the airplane to the ground.

The team found that if the ground was partially obscured by weather, their algorithms could even sometimes pick up and accentuate the faint contours of the target through the miasma — that is, they could see through clouds.

In 2011, when Gorgon Stare I was deployed to Afghanistan, the Air Force began providing footage from the operations to Livermore's engineers so that they could test Persistics on real battlefield data. How the Air Force managed to channel all those terabytes of video from each mission across 9,000 miles from Afghanistan to California, Vaidya refused to say, but she did note that it took less than a day.

Once the imagery arrived at Livermore, the team would run it through the processors and experiment with their automated tracking systems. This was not just lab work. If the software uncovered any actionable intelligence, the engineers would send the findings to Beale Air Force Base and a number of other units, where analysts would validate the intelligence and channel it back to Special Operations Command and intelligence units in the United States and in the field. (A Livermore spokesperson disputed a number of these details but declined to go into specifics; the Air Force does not discuss Gorgon Stare operations.)

Though Vaidya's researchers were accustomed to working on sensitive projects, it was unusual to have such a direct hand in life-or-death missions on the battlefield. They tracked vehicles from points of interest and explored patterns in the traffic. They watched soldiers conducting raids and studied how the beat of a city changed in the aftermath of major events like bombings. They hunted for IEDs and their emplacers. In some cases, they saw people die on the screen.

Vaidya told me that she found the experience thrilling. "Here were the guys at Livermore, basic researchers, seeing, 'Oh my God, they just shot this guy, and I have imagery that I need to mine to be able to figure out when he planted the IED.' How cool is that?" She grew excited. "A kid that just came out of graduate school!"

At the height of the program, Vaidya was managing about 30 engineers. The researchers with no security clearance worked on the algorithms without ever seeing even a single frame of surveillance footage. Midlevel engineers with secret security clearances would wrangle the data, while senior Livermore engineers like Vaidya who held top-secret security clearances would conduct the analyses based on the tracks the algorithms turned out.

By the time Persistics was drawn down, in 2014, the Livermore team had processed Gorgon Stare imagery from more than 2,000 operations in Afghanistan. The software was transferred to Sierra Nevada, the defense contractor that built Gorgon Stare, and Vaidya moved on to a new project applying similar automated analysis techniques to classified intelligence collected from North Korea.

Persistics was not the only program of its kind at the time. DARPA and Lincoln Laboratory both built similar tools in the mid- and late 2000s, as did several contractors. These early efforts were not always embraced. Some hardened commanders had no patience for the sophisticated analytical tools developed for programs like Persistics, preferring to rely on the human eyes of their trusted soldiers over an algorithm developed in a faraway laboratory by scientists who had never set foot in battle. Though Constant Hawk had a watchbox feature, the analysts who worked on the imagery rarely used it.

But as the programs have improved, they have also grown in popularity. As of 2018 there were at least 14 "state of the art" known detection and tracking programs in existence, most of which were developed with funding from either the Pentagon or the intelligence community. All military wide-area-surveillance systems on sale as of this writing have automated tracking features. Even Gorgon Stare's teenage analysts may not be so bored anymore; in 2017 the Air Force spent several million dollars integrating a new, unnamed algorithmic processing system aboard its entire fleet (more on that in chapter 8).

Beyond simply helping you follow targets more efficiently, a reliable tracking device could enable a number of sophisticated analytical tricks, the kinds of things that only a seasoned analyst might otherwise be capable of. Say you are watching a single node. "Is that building a weapons cache?" said Vaidya. "Or is it a meeting point for a bunch of bad guys? Or is it just a pub?" The software can track how many potential targets enter and leave the space, and at what times — patterns that may point to the building's function. A similar program developed by the Air Force Research Laboratory can track, according to the engineers who built it, every visitor to a known "drug house" in order to map out a network of users and dealers.

Building on these ideas, programs like DARPA's Wide-Area Network Detection sought to automate the type of attack-the-network analysis that the CIA would previously carry out by hand with Constant Hawk footage from Iraq. In one particularly revealing research paper discussing the concept, a notional schematic of the enemy is plotted on a satellite view of a US city. On the eastern edge of the city, there is the tipoff point, which is connected by a track to a known cache site, which is, in turn, connected by another track to a "suspect neighborhood," and so on. It looks like a detective's bulletin board mapping out the members of a mob family.

Tracking and detection algorithms can also allow one to extract much more information out of each pixel, reducing the need for large, expensive, high-resolution cameras. The camera that surveyed an area the size of Wales in 55 hours had only nine megapixels. For the human observer looking at its grainy footage, it is all but impossible to distinguish the people and rafts floating in the water from the whitecaps and multitudinous specks of light reflecting off the surface of the sea, but that doesn't matter, as the camera comes paired with a sophisticated motion-tracking software developed by the Australian firm Sentient Vision,* which detects all potential targets of interest — such as pirate dinghies or shipwreck survivors — automatically. The US Coast Guard is interested in buying the device for search-and-rescue missions, and it isn't hard to see why.

Reading Between the Lines

Remember — simply finding a target and tracking it from point A to point B is only half the battle. Whenever a brute-force analyst working on tracking vehicles with Gorgon Stare sees something that may be actionable — "a possible," analysts call it — the finding is forwarded to the more experienced team members for further analysis. This is where the second main task in imagery analysis comes in: "signification," the process of decoding the significance of what you are seeing. It is what you are doing when you notice a slick of oil on your

---

* The same company that is building the wide-area sensors for Australia's armed Reapers.

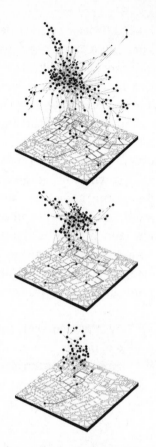

An illustration showing how vehicle tracks in wide-area surveillance are used to map the "social network graphs" of adversary organizations. Each line on the map represents the route of a suspect vehicle that was tracked from one node to another. *Emily Wissemann*

driveway and deduce that your car is leaking, or when a CIA analyst watches a man visiting a known terrorist safe house and surmises that he, too, may be a terrorist. It's what Sherlock Holmes is talking about when he tells Watson, "You can see everything. You fail, however, to reason from what you see."

Because a smart adversary will never lay all his cards on the table, most advanced imagery analysis requires one to reason from what one sees. During the Cold War, the Russian nuclear-submarine facility at Severodvinsk carefully timed its launches to avoid detection by US satellites. But the CIA's photographic interpreters—masters of signification — didn't need to see a submarine itself to know when something was afoot. They could predict each imminent deployment by merely noting the presence of keel blocks in the facility's launch basin — a sign as telling as an oil slick on a driveway. By counting and measuring the blocks, the analysts could even figure out exactly what kind of submarine was about to slip into the Barents Sea.

Such analytical gambits were standard practice at the CIA's imagery-analysis headquarters at the Navy Yard in downtown Washington, DC. A senior analyst who worked on the Severodvinsk case later boasted that his team could spot a new missile silo site "before they ever dig a shovel of dirt": a fence cutting across a remote forest usually meant that a new launch site was about to go up; a series of flattened pegs in the ground indicated a survey project. A good analyst could look at a column of tanks and decode the adversary's battle plans just by noting how far the vehicles were dispersed from one another.

None of this is easy. Any nineteen-year-old with basic analysis training can spot a row of keel blocks or a vehicle traveling between two nodes — even a fairly simple algorithm can follow a pixel as it moves across a screen. But if you want to know what it all *means*, you need an expert — or a very advanced computer.

8

Ghost in the Stare

ONE DAY IN the spring of 2008, Colonel John Montgomery walked into a ground-control station at Creech Air Force Base in Nevada for his regular shift flying a Predator over Iraq. As vice wing commander of the 432nd Expeditionary Wing, Montgomery kept a busy schedule, but he made a point of flying every week. The mission that day was an open patrol over Sadr City, a densely populated neighborhood in northeastern Baghdad. The Army was engaged in heavy fighting in the neighborhood, which had recently become a hotbed of insurgent activity, and Montgomery's squadron had been watching the area for weeks.

As Montgomery settled into his seat, his sensor operator, who was staying on from the previous shift, turned to him. "There's something wrong in this city," he said. "I don't know what it is, but things just don't feel right to me." Montgomery sensed it, too. "It was a vibe. It just wasn't right," he later told me of the mission. Montgomery's crew had become so accustomed to watching the streets of Sadr City that they understood the rhythm of the neighborhood; they even committed to memory the exact spots where local women hung laundry from their balconies. When something was off, it was obvious.

Fifteen minutes into the shift, the sensor operator pointed to a man on the screen. "This guy does not make sense to me," he told Montgomery. The man was wearing a suit, and he was speaking on a cell phone. From 15,000 feet, he wouldn't have appeared particularly unusual or suspicious to the untrained eye. But the sensor operator was an experienced airman and Montgomery trusted his instincts. He agreed to drop the planned patrol. On the basis of a hunch, the man in the suit became a target.

For more than three hours as the Predator orbited overhead, the man didn't once set foot inside a building. He seemed to be walking aimlessly, at times strolling down the middle of busy roadways. He kept his cell phone to his ear the entire time.

Eventually, the man made his way into a quiet side street and a Toyota pickup pulled into the frame. Three men emerged and, together with the man in the suit, the group took a mortar tube out of the truck's flatbed tray and fired two shots toward a nearby US base. After dumping the barrel in an abandoned lot, the three men got back into the Toyota and drove away, and the man in the suit went on walking as though nothing had happened.

An intelligence team was dispatched to follow the Toyota, and a second team crossed the neighborhood to retrieve the mortar. The Predator crew continued following the man in the suit, who disappeared into a house a few blocks away. Montgomery said that he met his Maker shortly thereafter.

Even a simple tracking system could have followed the man in the suit around the city, but it would not have been able to tell that he was, in fact, a member of the insurgency. In the operation over Sadr City, the mission hinged on a decision based on subtle cues, lots of experience, and a heavy dose of intuition. Surely a computer wouldn't be capable of *that*.

Unknown Unknowns

In January 2017 I drove to the headquarters of Kitware, a few miles north of Albany, New York, to find out how the software-development company was creating computers seemingly possessed of all the faculties of an experience human analyst.

Kitware has an outsize influence in the world of software development. One of the company's specialties is what's known as visualization. Prior to joining the Angel Fire project in 2005, it worked with Los Alamos, building software to visualize simulated nuclear explosions. (The US government, as a signatory to the Comprehensive Nuclear Test-Ban Treaty, has not been able to conduct actual nuclear trials since the treaty went into force in 1992, so it has had to test its warheads virtually.) Combining a vast range of data points, including

the megatonnage of the nuclear device, the location of the hypothetical blast, and the weather, the simulations were so detailed and complex, it would often take a Los Alamos supercomputer several weeks
to process a single imaginary detonation.

When Angel Fire was transferred to the Air Force Research Lab
and Kitware was dropped from the program, Charles Law and Bill
Hoffman, the company's founders, felt that the Air Force had missed
an opportunity. Not only was the system unfinished and glitchy, it
was also wasting perfectly good data that could be mined for intelligence.

Determined that their work on the concept not end with Angel
Fire, the company formed a computer-vision group that would focus, among other areas, on developing automated aerial-surveillance
software. To lead the program, Hoffman and Law picked Anthony
Hoogs, a soft-spoken engineer with a PhD in computer vision from
the University of Pennsylvania who had previously worked on intelligence programs for Lockheed Martin and GE Global Research.

At the time, wide-area surveillance was based on a "right of boom"
concept of operations, meaning that an investigation into a network
could begin only after an explosion had occurred, since the explosion
provided the initial lead in tracking potential targets and, hopefully,
preventing future attacks. When John Marion pitched the initial concept for wide-area surveillance around the Pentagon, officials would
say, "Well, we don't want to start with the IED going off to figure out
who all is involved. We don't want the IED to go off at all. How are
you going to do that?" The whole point of the vast, Manhattan Project–like effort was to get "left of boom."

But a single 10-hour Gorgon Stare mission generates 65 trillion
pixels of information. To borrow a popular analogy in the intelligence community, finding an insurgent without an explosion to tell
you where to begin your search is like trying to find a needle in a hay-
stack, in a field full of haystacks, in a county full of hayfields.

One day Hoogs was watching a sample of video footage from a
wide-area camera when he happened to spot activity on the screen
that to him seemed suspicious enough. He wouldn't tell me what
it was but he said that, had the footage been live video, it probably
would have merited some kind of action. Understanding that the ac-

tion was suspicious required signification. Hoogs wasn't watching a man literally planting a bomb or firing a mortar but rather something that suggested such an activity was about to occur. And yet the activity itself was simple and obvious enough to be reliably identified by a computer.

This led to a revelation. An all-seeing camera flying over Baghdad or Kabul would capture all manner of activities that the defense and intelligence communities would want to take a closer look at — meetings between combatants, men loading IEDs into trucks, emplacers digging holes by the side of the road — all things that, without an explosion, would go unseen until it was too late. If it was possible to automatically spot indicators of an imminent attack, could human operators be alerted with enough lead time to intervene? Simply put, could a crime be stopped before it happened?

Hoogs's idea for using computer vision to preemptively detect criminal acts tapped into an increasingly popular theory of spycraft. Known as activity-based intelligence (ABI), it had grown out of a series of special-operations missions against "dynamic targets" (another term for people with ignition keys) in Iraq and Afghanistan. Like the man in the suit, these adversaries bore none of the traditional markers of a belligerent. They wore no uniforms, they didn't drive tanks; typically they didn't even carry firearms. Their only weapon might be a cell phone. They blended seamlessly into the local population.

Practitioners of ABI regularly invoke former secretary of defense Donald Rumsfeld's widely pilloried remark about "unknown unknowns," made during a press briefing on Saddam Hussein's links to terrorist groups. This is because, when seeking out dynamic insurgents, you often don't know exactly what you're looking for. Letitia Long, a former director of the National Geospatial-Intelligence Agency, described the task as "like looking in a global ocean for an object that might or might not be a fish. It might be anything and it might be important, but at first, we are not sure it even exists."*

---

* The official literature on ABI is full of esoteric aphorisms that wouldn't feel out of place in a volume of Buddhist koans: "Everything happens somewhere," write the authors of a 2017 paper describing ABI techniques. "Nothing can be in two places at once. Nothing can be nowhere." (Patrick Biltgen et al., "Activity-Based Intelligence:

According to a report by a senior Pentagon intelligence official that has not been publicly released, ABI often looks for "nontraditional" markers of participation in an armed group, including a number of indicators that might not, on any level, appear to have anything to do with being a combatant. These include name, gender, age, weight, religion, skills, biometrics, values, race, email address, and even personality traits.[*]

The most obvious and straightforward marker of an unknown unknown is an individual's activity, hence the theory's name. Innocent civilians don't plant IEDs or fire mortars. Nor do they drive between known safe houses in the dead of night, or place calls to numerous known insurgent leaders. As a DARPA program manager specializing in intelligence analysis put it in 2009, "Bad guys do bad things."

The activity that Hoogs discovered in the test footage was precisely this kind of unknown unknown: nobody knew that it was happening, and nobody knew the identities of those involved, but it was clearly a bad thing.

The principal challenge of building a computer-vision algorithm for detecting such activities, and various others like it, was that it would have to be flexible enough to account for the messiness of war. On a pixel-by-pixel basis, there are a million ways to commit a crime, and the computer would have to recognize each variation of that single suspicious action. When Kitware, along with researchers from the University of Maryland, Berkeley, and Georgia Tech, set out in 2008 to test a preliminary detection system, using footage from an actual war wasn't an option, as some of the researchers lacked a security clearance. Instead, they set their algorithms loose on what they considered the next closest thing, a season of Georgia Tech football game tapes.

Like combat, football is both organized and chaotic. Each play is an encounter between two meticulously thought-out strategies that

Understanding Patterns-of-Life," in US Geospatial Intelligence Foundation, *The State and Future of GEOINT 2017*.)
[*] ABI operates on logic similar to that of the controversial "signature strike," by which drones are used to strike targets whose exact identity is not known, but whose actions and/or associations suggest participation in an enemy organization.

may result in an entirely unpredictable outcome. Like the insurgent groups in Baghdad, Hoogs said, football players engage in active deception to try to mislead their opponents. The goal was to build a motion and tracking algorithm that, when applied to the football video, could read the intentions of the offensive line before it actually executed its plan.

Since the engineers themselves didn't know the difference between a "sweep," a "bootleg," and a "quarterback sneak," they enlisted a Georgia Tech graduate student who had briefly been drafted as a defensive back for the Green Bay Packers to help guide the algorithm to look for the early telltale markers of a range of particular tactics. By the end of the yearlong initiative, the computer was able to recognize seven types of play within just three seconds of the hike. To be sure, the human defensive back could discern a play within two to two and a half seconds, but the algorithm was still much faster than the average viewer at home.

A U-Turn at 3:00 a.m.

Shortly after completing the football project, Kitware won an $11 million contract to develop a computer-vision system for DARPA that would autonomously identify certain activities in grainy narrow-field footage. The idea was to single out suspects—people like the man in the suit in Sadr City—who were otherwise impossible to distinguish from the general population. The manager of the program was the same official who said that bad guys do bad things.

This initiative enabled Kitware to develop a number of tools that Hoogs believed could be applied to his wide-area-surveillance idea, and he pitched DARPA for a seed grant to carry out some developmental work on his concept. Hoogs and his team spoke to a number of airborne-intelligence analysts about the types of activities that —like the actions of an offensive line—presage an adversary's next move or an imminent attack. After a few months of work, Hoogs demonstrated that his algorithms could pick up a handful of these behaviors with a high level of accuracy. This led to yet more DARPA funding. According to Hoogs, on the basis of the demonstration DARPA created a program to build an artificially intelligent all-seeing eye. Officials named it the Persistent Stare Exploitation and Analysis System — PerSEAS. Perseus is the Greek mythological hero who kills the Gorgon Medusa.

According to a document outlining the program, the goal was to build something that could "identify threats early enough . . . to make a difference," including things like "driving behaviors occurring before the detonation of a suicidal vehicle" and events that "highlight 'patterns-of-life' associated with a variety of network types, including social, political, regional, economic or military networks." (The document noted that any software capable of "face recognition, gait recognition, human identification or any form of biometrics" was strictly prohibited.)

It didn't matter if the computer couldn't catch everything. As long as it caught more "unknown unknowns" than one would find with human eyes alone, and as long as it didn't produce *too* many false negatives, according to Hoogs even a 10 or 15 percent discovery rate on otherwise undetected activities would be worthwhile.

In 2010 DARPA awarded Kitware a $13.8 million contract to lead the work on PerSEAS. Hoogs assembled dozens of academics and researchers from four major defense contractors and six universities, including Berkeley, Columbia, Rensselaer Polytechnic Institute, and the University of Maryland.

As a matter of policy, most of the behaviors that Kitware's computers were trained to identify remain secret — the intelligence community doesn't want insurgent networks finding out which specific behaviors might tip their proverbial hand, but a few have slipped out. Vehicular events of interest, listed in a nonpublic Kitware presentation, include "keep distance far" (when one car follows another from a distance), "keep distance close" (when a vehicle follows closely behind another), "passing" (when a vehicle overtakes another), "flip-flop driving" (when two cars pass each other repeatedly), "approach" (when a car drives toward a fixed location), "retreat" (when it drives away), "parallel driving" (when two cars drive side by side), "dropping off" (when a vehicle comes to a stop and leaves a passenger on the curb), "aimless driving," and "meeting" (two cars coming to the same spot within a short time).

When the team turned its honed software on a wide-area feed of a

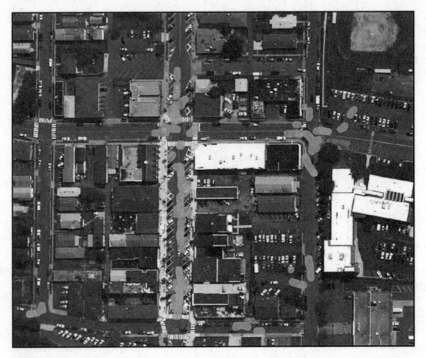

The results of a test in which Kitware's activity-detection system looked for "vehicle-stalking-vehicle" events (when a car waits for another car to vacate a parking spot, then takes the spot) in a clip of WAMI footage. The shaded areas indicate locations of all vehicle-stalking events detected by the computer. *This material is based upon work supported by the Defense Advanced Research Projects Agency (DARPA) under contract no. W91CRB-10-C-0098. The views and conclusions contained in this document are those of the authors and should not be interpreted as representing the official policies, either expressly or implied, of DARPA or the US Government. Approved for public release; distribution unlimited. © 11/06/18 Kitware, Inc. All rights reserved.*

large city, the results were impressive. A single data set might include 10 million vehicle tracks and 100 million events. At the click of a button, the computer could flag the exact locations on the map where every particular event of interest had occurred. It was as though you could snap your fingers and see every car that had left a parking spot in your neighborhood.*

---

* As a quick aside, some of these same techniques are also applicable to social media, which have become a fruitful source of intelligence for the Pentagon and the intelligence communities. The Defense Intelligence Agency uses a computer-vision

One type of activity that is of particular interest to groups that hunt insurgents is the U-turn. In the early years of the Iraq war, analysts began to notice that if targets suspected that their vehicle was being followed, they would often make a U-turn and head back toward the chase vehicle. It became a telltale sign that the individual being followed might have something to hide.

As early as 2005, engineers at Livermore and, later, Logos had begun developing algorithms to automatically detect U-turns. But they soon ran into trouble. U-turns, like most of the other vehicular indicators of suspicious activity, were common in Baghdad. Analysts wouldn't have time to investigate every algorithmically detected drop-off or instance of "parallel driving" on a given shift; if they did, they'd probably miss much more important activities. And this was if the algorithm always successfully identified what it was looking for, which it didn't. At one point in a Kitware test video I watched, a "meeting" that the algorithm detected was merely two cars parking next to each other in a parking lot.

The PerSEAS engineers began coding the software to recognize when a single vehicle or set of vehicles engaged in a combination of such activities, something that was much less common and much more indicative of insurgent activity. Though analysts would ignore a single isolated U-turn, they would be very interested in a U-turn followed by a 10-minute stop at the side of a busy road, a clear indicator, according to John Marion, the Logos engineer, of a nefarious activity.

But even a U-turn followed by a lengthy stop could be fairly common, and it might not be basis enough to rationally suspect a vehicle's occupants of being terrorists. That is, unless you look at the activity in context. A man in a suit talking on a cell phone for hours wouldn't necessarily be a suspicious sight in Lower Manhattan, but in the middle of war-torn Sadr City, the visual was a clear anomaly. Hoogs pointed out that while it is very common for two cars to stop

---

system to discern significant threats in videos posted online by insurgent and terrorist groups, while IARPA has funded Kitware to develop software that can autonomously single out YouTube videos featuring certain activities of interest with remarkable precision. In one test, it identified videos of "flash mobs" with a 74.3 percent accuracy rate out of an archive of 26,000 clips.

next to each other in a busy parking lot during the day, the same activity in an empty lot at 3:00 a.m. is much less common. Likewise, cars stop at intersections all the time, but if a car stops at the side of a busy highway, it's possible that its occupants are planting a bomb (it's also possible, Hoogs conceded, that the car has simply broken down).

Like the track, the "anomaly" is a fundamental element in spycraft. At GE Global Research, Hoogs had worked on a system, also for DARPA, that sought out anomalies in maritime traffic data to detect nefarious shipping activities (ships with innocuous cargo travel along scheduled, predictable routes; ships with illicit cargo do not). Credit card companies employ a similar strategy to detect fraudulent transactions, which is why banks will often block your card if you attempt to make a withdrawal in a foreign country.

Using techniques known as cluster analysis, Kitware built an algorithm that segmented the area under surveillance into patches and generated a "normalcy model" of the routine activity in each patch. Watching a patch containing an intersection, for instance, the algorithm would measure the average speed, size, and type of passing vehicles, how often vehicles stopped, and the frequency of U-turns. With a model established, the computer would then single out only those activities that broke significantly from the model.

Whereas it took months if not years of experience for Montgomery to learn how to recognize when something was amiss in Sadr City, these systems would require only a couple of hours to learn the difference between what's normal and what's not in any given area. Using similar techniques, the defense contractor Harris has built a traffic-speed normalcy model for the entire city of Rochester, New York; the program can detect when a car is traveling above the average speed in a particular location or when a vehicle stops in an area where most cars tend to move quickly.

When I was driving to meet Hoogs at the Kitware office in the winter of 2017, my phone, which I had been using for directions, had run out of battery. After making several sudden lane changes on the highway, I pulled off at an exit and stopped at a gas station on the outskirts of Schenectady to find a charger. When I returned to the road, twenty minutes later, I had to make an awkward U-turn and retrace my route along that same highway for about five miles.

As Hoogs described the PerSEAS project, it occurred to me that some of my behavior on the way to the interview had matched the types of activities that Kitware's computer-vision systems were designed to find. Not only had I been making all sorts of turns and stops during my unintentional detour, I now realized, but I had also been doing so in a quiet upstate New York suburb where that type of driving was, in all likelihood, quite rare.

I asked Hoogs how the computer might have responded if it had been watching me. From the sound of it, Hoogs said, my driving was exactly the kind of behavior that would have alerted the system. Other intelligence specialists and engineers confirmed to me that my actions would have definitely raised eyebrows among both human and robotic intelligence analysts in the field.

Threat Level: 99 Percent

A couple of years into the PerSEAS program, the DARPA manager who had been administering the effort left the agency. According to Hoogs, his replacement reoriented PerSEAS to focus on vehicle tracking. The autonomous tracking system that Kitware produced, called the Kitware Image and Video Exploitation and Retrieval system, or KWIVER, for short, was deployed to the battlefield, though Hoogs would not say where, or by whom. In 2014 the Air Force Research Laboratory described KWIVER as the field's "state of the art" system.

At the time of my visit, Kitware's work on detection algorithms was ongoing. The company has been awarded a number of Pentagon grants from government agencies and labs, including the Air Force Research Laboratory, to develop the software further. It also built event-detection software under a broad program run by the Office of the Secretary of Defense called Data to Decisions, which sought to apply advanced automated analytics to incoming feeds from a range of surveillance systems. All told, Kitware has received almost $40 million in government contracts for this work. From 2009 to 2013 it consistently ranked among America's 5,000 fastest-growing companies.

The potential of an effective behavior-detection system like the one developed by Kitware is lost on nobody in the intelligence community, and a number of other groups and contractors have developed similar software to root out unknown unknowns. When combined with the type of tracking capabilities described in the previous chapter, it could automatically turn a baffling and overwhelming mass of data into detailed actionable intelligence about not just individual actors, but complex networks.

A set of researchers, from Kitware, Los Alamos, and the University of New Mexico, have suggested that a finely tuned anomaly-detection system could, for example, calculate the probability that two anomalies occurring in the same neighborhood represent a coordinated effort involving multiple parties in a single organization, just like how McNutt's analysts identified cartel-like force-protection patterns in Baltimore, and even identify the places in a city where they are likely to regroup so that forces on the ground can be waiting for them when they arrive.

One particularly revered company working in this area is Signal Innovations Group. Founded in 2004 by two engineers from Duke University, SIG applies sophisticated data analytics to a variety of aerial spying technologies, including sonar, radar, sensors for detecting IEDs, and WAMI. While neither of the company's founders responded to multiple requests for comment, we know from other sources that SIG's programs have already seen extensive use in the field, primarily for classified operations. The company has worked with the Office of Naval Research, the Air Force Research Laboratory, DARPA, the Army Research Office, and the Army Night Vision and Electronic Sensors Directorate, in addition to its undisclosed customers.

SIG has developed its own automated analysis system similar to Kitware's. And it's remarkable. A "Competition Sensitive"* Power-Point presentation created by the company in 2013 explains that the goal of the program is to take quadrillions — in other words, millions

---

* A designation used for documents that contain proprietary information.

of billions—of pixels of imagery and whittle the data down to a few hundred potential leads for human analysis.

Among other things, the document claims that SIG's software could calculate, based on airborne surveillance alone, the likelihood that a given location is a terrorist node. It appears that the company demonstrated this precise capability using data from one of the many secret WAMI test flights over American soil: the CIA-sponsored exercise over Lubbock, Texas.

The presentation indicates that to begin the process human analysts will mark known terrorist nodes on a map. The software will then "propagate" the threat across the city, pointing to new potential terrorist buildings by tracking vehicles from the original known nodes. The company claims that the software is capable of identifying the origin and destination of "a large percentage" of all vehicles moving in the surveilled area; one slide features a satellite map of Lubbock showing the paths of what appear to be thousands of tracked vehicles.

The system then moves on to subtler strategies to find further unknown unknowns. The presentation appears to suggest that the software tracked vehicles arriving at and departing from eleven different types of locations around Lubbock, including homes, workplaces, soccer fields, restaurants, and even barbershops. It then built a normalcy model for the number of visits that each type of location receives over the course of a typical day, similar to the normalcy models of vehicle speed generated by Kitware's software. The SIG software can then predict the "role" of an enigmatic location by matching its pattern of visitors to the arrival-and-departure models from known buildings. If vehicles congregate around a particular site between 4:00 p.m. and 8:00 p.m., the site might be another soccer field. If a building has a spike in visits at lunchtime and dinnertime similar to visitor patterns observed for a known restaurant, it could be surmised that the building is a restaurant.

In addition to matching vehicle activity patterns around unknown buildings to that of known restaurants, IED cells, and barbershops, SIG claimed that its software can also identify suspicious nodes when there is "anomalous building activity" that does not match the normal visitor patterns for that type of location—say, a restaurant that sees a spike in visits between 2:00 a.m. and 4:00 a.m.—or if a building receives visits by vehicles that have exhibited anomalous driving behaviors. Had such a system been watching me while I was driving through Schenectady, it might have flagged Kitware's office for receiving a visit from a suspicious vehicle.

In 2014 the defense contractor BAE acquired SIG for an undisclosed sum. BAE now sells software based on the same system that the company tested using the data from Lubbock. According to BAE marketing materials, it is capable of both tracking large numbers of vehicles simultaneously and identifying patterns and anomalies that might indicate the presence of suspicious groups: autonomously separating the good from the bad.

Since there are, presumably, no IED cells in Lubbock, the company appears to have used the city's medical industry as a stand-in for a terrorist network in its demonstration. In a second satellite map of the city, a doctor's office called West Texas Pediatrics (which is a real business in Lubbock) is marked with a label indicating a "threat probability" of 99 percent. A second business farther north, Medical Management Solutions (also real), has a threat probability of just 15 percent. Why SIG's software deemed West Texas Pediatrics so dangerous, the presentation doesn't specify.

A Little Deep Learning Is a Dangerous Thing

In early 2017, after more than a decade of government investment in laboratory experiments, a Pentagon task force concluded that advanced imagery-analysis algorithms "can perform at near-human levels." Following the task force's recommendation, DOD leadership launched Project Maven, a much-publicized and yet somewhat shrouded effort to take automated video analysis to war.

The goal of the program is simple: to get the best possible algorithms into the battlefield as quickly as possible. Like Big Safari, Project Maven deploys technologies when they're only 80 percent complete. Also known as the Algorithmic Warfare Cross-Functional Team, the program's first experiment, an automated analysis system

The Pentagon's official seal for the Algorithmic Warfare Cross-Functional Team, also known as Project Maven, which is looking to build artificially intelligent surveillance-imagery-analysis systems. *US Department of Defense*

capable of recognizing targets and discovering suspicious activities in soda-straw drone videos, was delivered to 10 intelligence units working on missions over Syria, Iraq, and a number of undisclosed African countries in late 2017.

Among the software's many features, analysts can select a target of interest and the software will assemble every existing clip of drone footage showing that same vehicle or individual spotted in previous missions. Other features are classified, though they can be fairly easily guessed. One of the contractors on the project, a computer-vision startup called Clarifai, sells software capable of analyzing a person's "age, gender and cultural appearance" in videos and photographs.

In its second "sprint," Project Maven turned to WAMI. By the end of 2018 the program will have deployed an "AI-based" analysis algorithm for Gorgon Stare, which will be available on the Pentagon's internal top-secret intranet systems, SIPRNET and JWICS. The

Air Force declined to tell me exactly what kinds of analysis this all-seeing AI would be capable of, but a Pentagon budget document released in early 2018 noted that it could perform target identification and recognition. In a presentation to members of the Royal Australian Air Force, the director of the program showed that an early prototype of the software was capable of instantaneously recognizing cars, trucks, people, and boats. He also demonstrated how an analyst could select an area of interest and the software would generate a total count of people or vehicles visible in the frame, and suggested that it would eventually be capable of more sophisticated tasks — the kind of things one might call signification.

If successful, Project Maven will open the door to a new era of artificially intelligent spycraft. In its budget request for 2019, the Office of the Secretary of Defense, which allocated a hefty $109 million to the effort, called it a "pathfinder" for the DOD's broader work in AI. Partners in the initiative include the US national laboratories and all 17 member organizations of the US intelligence community.

The main barrier to the widespread adoption of this technology earlier in its history was the fact that even the most advanced of these systems are not entirely glitch-free. In one 20-minute test, I noticed that Kitware's behavior-detection algorithm flagged as suspicious an intersection where a car pulled away from a stop sign and a second car replaced it from behind seconds later ("replacement" — when one car replaces another car — was one of the signatures that the software was designed to search for). In another, it flagged as an example of replacement a clip of a single car making a three-point turn.

No soldier would be able to trust a system that makes such glaring errors. In one test under the DARPA Mind's Eye program — which sought to build software that reliably recognizes 48 different human actions, including "have," "turn," "hit," "dig," "bounce," "run," "snatch," "throw," "touch," "exchange," "replace," and "flee" — a prototype developed by a team at Colorado State University was able to correctly identify in test surveillance video the action of a woman turning, "but," the researchers lamented, the software, which was supposed to detect whenever a figure on-screen was holding an object, "missed that [the woman] is carrying a bowl of fruit."

But such mistakes are becoming more rare. This is thanks in large

A screenshot from tests of a computer-vision program developed at Colorado State University. The computer was able to correctly identify a woman turning around, "but," the researchers lamented, "we missed that she is carrying a bowl of fruit." No soldier would be able to trust a system that makes such glaring errors. *Bruce A. Draper, Mind's Eye PI Presentation, http://www.cs.colostate.edu/~draper.*

part to recent advances in computing in the sphere of machine learning, a technique that involves sharpening an algorithm's capabilities on vast training sets.

The profound impact of machine learning, as well as its cousin, deep learning, which involves a similar training process using an architecture modeled on the brain's neural networks, is already apparent in modern life. Take YouTube's exquisitely effective video-recommendation sidebar, which employs a cluster-analysis technique similar to Kitware's anomaly-detection software. Just as Kitware's software determines that a U-turn on a highway might indicate that something's not right, YouTube can predict that a viewer who searches for a video of a Huey helicopter will probably also be interested in a video about a nuclear submarine or a clip from *Apocalypse Now*. The reason that Kitware's software makes regular mistakes while a visit to YouTube rarely ends with one video is that YouTube's software has been trained on over 100 billion examples of viewer behavior. It is for similar reasons that Facebook's facial-recognition system (which is trained on masses of selfies) is far more accurate than the FBI's (which is trained on much smaller data sets).

In the field of automated intelligence, even comparatively little learning, it turns out, goes a long way. The impact of deep learning, in particular, a Jason group study concluded in 2016, "is nothing short of revolutionary." When paired with a deep-learning system trained on ImageNet — a training database containing 14 million annotated images of everything from horses to famous buildings — the Lincoln Laboratory's automated WAMI-analysis software's false-alarm rates dropped to almost zero. A deep-learning-based tracking and detection system unveiled in 2017 by a team from the University of Central Florida reportedly outperformed 13 other non-deep-learning programs by up to 50 percent. Much of Project Maven's early success likely comes down to the fact that its analysis software is trained on over 1 million tagged images.

Skeptics of these efforts say that even such extensive "training" in a laboratory could never fully prepare an algorithm for the unique chaos of war — nobody becomes a true expert without having been in the fight. The repeated mistakes that Kitware's software made in hunting for "replacements" were the kind of errors that a human would only make once, as long as the error was pointed out to him or her. Groups like the Lincoln Laboratory, the Air Force Research Laboratory, and the Wide-Area Surveillance Laboratory at the University of Dayton, among others, are therefore building "active learning" features into their software, which allow analysts to correct the computers' work midmission. Under Project Maven, when the system misidentifies an object or activity on the ground, analysts can click a "train AI" button and the algorithm will remember not to make the same mistake again in the future.

Likewise, when the computer is right, the analysts affirm it. This way, the computer builds an understanding over time of what works and what doesn't. (If you've ever had to identify road signs or street numbers in a CAPTCHA test while filling out an online form, you've participated in a similar kind of human-supervised learning program for Google Maps; by typing in what you read on the street sign, you are essentially confirming the computer's work, thereby helping to sharpen Google's AI even as you verify that you are "not a robot.") The longer the system is in operation, the better it will get at its job.

These systems will also become more versatile. "A pilot trained in Philadelphia should be able to fly the plane in Phoenix," explained Vijayan Asari, from the University of Dayton's lab; computer programs should be no different. If an operation moves from, say, an urban battlefield to a rural one, active-learning algorithms will quickly adapt to the new environment, with only minimal guidance from a human operator.

Some programs will even be capable of learning new tricks on the fly. Analysts using Project Maven software, for example, can teach their systems to recognize entirely new kinds of events that they were never trained for. In one example provided by the program's director, an analyst could label an unfamiliar scene as "an emergency" and identify two vehicles in the frame as a fire truck and an ambulance. The next time a similar incident occurs and a fire truck rolls into view, the computer sees it on its own, and knows exactly what it means.

Even people with firsthand experience in the jobs that these algorithms are designed to replace find it hard to deny the power of the recent advances in machine learning. When I spoke to Colonel John Montgomery, the drone pilot who tracked the man in the suit, in December 2016, he still wanted to believe that the human touch will always be necessary in those kinds of missions — but with the advent of machine learning, he was no longer so sure.

The day before we spoke, Montgomery had watched a YouTube video of a robot learning how to play ball-in-a-cup. At first, the robot, which employs a form of machine learning, is as hopeless as a toddler. At the 70th attempt, the ball hits the rim of the cup. After 90 attempts, the ball is consistently hitting the rim. On the 100th attempt, the robot makes it. "Well, after that," Montgomery said, "It never missed. Dunk, dunk, dunk, dunk."

Silicon Valley Enters the Fray

The undisputed leaders in the type of machine learning necessary to train computers to match a human analyst are, of course, the firms based in Silicon Valley. This is in no small part because of the tech industry's ability to poach talent from other sectors, including the defense and intelligence world. Much of the team from Lawrence Livermore's Persistics project, for example, left to take jobs at Google, YouTube, and Facebook, among other firms. According to Sheila Vaidya, this was the main reason the laboratory had to shutter the effort in 2014.

The Pentagon is eager to put those minds and technologies back to work on its airborne-intelligence operations, and Silicon Valley has signaled that it is willing to cooperate. In the fall of 2017 the Office of the Secretary of Defense selected Google as a partner in Project Maven. Under the agreement, the Pentagon would use the internet giant's TensorFlow artificial intelligence system to power its various analysis programs. Internal company emails show that senior Google executives predicted the deal might eventually generate up to a quarter of a billion dollars for the company.

When the partnership was revealed in March 2018, the story was circulated widely. Google issued a statement explaining that the company's tools would be used only for "nonoffensive" activities. Even so, more than 3,000 Alphabet employees signed a petition urging the company's CEO, Sundar Pichai, to call off the partnership, which he did shortly thereafter.

Many observers of the scandal were shocked by the news of the collaboration, coming from a company that had for years lived by the motto "Don't be evil." But this was not the first time Google has assisted in an airborne automated spycraft program. In around 2013 Google signed a cooperative research and development agreement with the Air Force Research Laboratory focused on data-processing technologies for, among other applications, aerial surveillance.

A CRADA is a partnership between a government agency and a company or university intended to foster the development of private sector technologies and products that the sponsoring government institution wishes to eventually deploy in active operations. As a result of the AFRL-Google project, Air Force engineers developed what a webpage posted by the Office of the Secretary of Defense describes as a "revolutionary" prototype for automated pattern-of-life analysis in wide-area-surveillance footage.

This was not a "nonoffensive" technology. Pattern-of-life analysis, you'll recall, is the process of studying an individual's daily activities in detail from above. It is an integral step in the process leading up to an airstrike. An Air Force spokesperson I contacted declined to provide further details about the agreement.

There are other signs that Google's involvement in the defense world was more extensive than previously thought. In its 2019 budget request, the Special Operations Command noted that it needed $4.5 million to purchase a number of computing services, including, it notes, TensorFlow, for a "big data analytics" program. The Air Force spokesperson would not confirm or deny whether the AFRL-Google CRADA was the only case of Google participating in an Air Force project and noted that the service "will continue to partner with industry and academia pursuing new and emergent technologies to enhance our decision-making."

Following the Project Maven controversy, Google's leadership did appear to temper, at least temporarily, its earlier zeal for DOD dollars. In October 2018 the company withdrew from the competition for a Pentagon cloud computing program known as JEDI (for Joint Enterprise Defense Infrastructure) that is likely to be worth in excess of $10 billion. According to a statement, the decision to abandon the program was based in part on the fact that the company "couldn't be assured that it would align with our AI Principles."

But there are still plenty of others in Silicon Valley with fewer compunctions about doing business with the military. "If big tech companies are going to turn their back on the US Department of Defense, this country is going to be in trouble," Amazon's CEO Jeff Bezos said at an event the week after Google's announcement, confirming that Amazon would remain committed to both JEDI and the military as a whole. Bezos's words were likely music to many ears within the Pentagon: the company that you may or may not have used to order this very book already provides all 17 intelligence agencies with a cloud computing system optimized for automated analytics under a $600 million contract awarded in 2013.*

---

* The cloud, which gives paying customers access to some of the most formidable computing power on the planet, is likely to have a particularly profound impact on the effectiveness of automated surveillance systems. The Air Force Research Laboratory found that using Amazon's cloud service dramatically improved the speed and accuracy of automated vehicle detection and tracking in a 140-megapixel video of Columbus, Ohio.

A week and a half after Bezos's speech, Brad Smith, Microsoft's president, announced in a company blog post that Microsoft, too, would stay in the running for both JEDI and other DOD technology ventures, including some, he acknowledged, that raised troubling questions: "We are not going to withdraw from the future."

Looking around at the future that Silicon Valley has already made for us — a future where you can order a book on a smartphone and then, with a flick of the finger, access thousands of hours of videos of helicopters and submarines that an artificially intelligent computer, somewhere, has decreed you'd want to watch — the tech world's growing closeness with the defense and intelligence communities is a bit of a heart-stopping prospect. Headlines that would otherwise seem benign, even welcome — for instance, "Finally: An App That Can Identify the Animal You Saw on Your Hike," or "Google Uses AI to Find Your Fine-Art Doppelgänger" — take on a sinister new meaning. All of these marvels could be co-opted for surveillance. In one way or another, most of them will.

9

New Dimensions

As SURVEILLANCE BECOMES smarter, it is finding its way into entirely new domains of our lives, from satellites up in the exosphere all the way down to the CCTV cameras on Main Street. Ultimately, there will be few corners of our planet that are not watched over, somewhere and somehow, by an all-seeing eye.

Take present-day CCTV cameras that blanket many modern cities. Just like Predator drones, they suffer from the soda-straw problem. In heavily secured areas like airports and casinos, the only way to ensure that nothing is missed is to install large numbers of cameras throughout.

One solution to the narrow-field view is a bigger camera. The first truly all-seeing CCTV was the Imaging System for Immersive Surveillance, a single CCTV unit made up of 48 individual imagers. Developed by the MIT Lincoln Laboratory and the Pacific Northwestern National Laboratory at the behest of the Department of Homeland Security, it was first tested in 2009 inside Terminal A of Boston's Logan Airport. With 240 megapixels, it was capable of reading the name on a boarding pass from up to 200 feet away.

Operators could pan, tilt, and zoom through the footage while the camera continued recording the entire space. Tracking algorithms allowed the DHS to keep tabs on particular individuals as they moved through the building, even during peak times when the images were filled with considerable clutter. Investigators could track through the footage as far back as 30 days. "You don't miss anything," said Bill Ross, the Lincoln Laboratory engineer.

After the initial tests at Logan, in 2014, the Boston Police Department installed a series of the cameras near the finish line of the Boston Marathon, and the Secret Service appears to have deployed one to the area around the White House. In 2016 MIT transferred the technology to Consolidated Resource Imaging, the wide-area-surveillance contractor, and the company sold its first two units in early 2017. In 2018 the company installed a unit at the CenturyLink Field stadium in Seattle.

Soon after beginning work on the Immersive Surveillance system, the Lincoln Laboratory began developing an even more remarkable ground camera called WISP-360, an infrared system built around the computer-in-every-pixel technology that can generate massive images with relatively small fast-spinning cameras and track speeding bullets. WISP-360 can create a full 360-degree hemispheric image composed of hundreds of millions of pixels every second, allowing one to see several miles in every direction, including in the sky. It operates much like a radar, except that it creates beautifully crisp black-and-white video. The exact specifications of the camera are classified, but when the Lincoln Laboratory flew the same device over Boston from 9,000 feet in the air, it was powerful enough to see individual cars driving around the MIT campus.

In 2012 the US Army began installing WISP-360 at a number of forward-operating facilities in active war zones. After ISIS and other nonstate groups started experimenting with hobby drones packed with explosives that could be used as rudimentary cruise missiles, demand for WISP-360 — which, thanks to its broad coverage and intelligent pixels, is capable of detecting small commercial drones — exploded. To accelerate production, in 2018 Ross and several of his team members spun off a private company, Copious Imaging, which will work to further commercialize the cameras, and continue developing the underlying computer-in-each-pixel technology.

At home, too, demand for ground-based all-seeing eyes is expanding. In early 2017 Customs and Border Protection outlined a plan to install over 200 new surveillance towers along the southern border to supplement the 200 that are already there. According to a solicitation for proposals, CBP was looking for a surveillance device that could watch a full 360-degree panorama and detect an average adult from up to 7.5 miles in the desert (or from up to 3 miles in a city).

The Israeli firm Elbit Systems has been pushing its two wide-area surveillance cameras, the GroundEye and the SupervisIR, for the CBP contracts, as well as for surveillance programs in "populated areas." Other firms have unveiled large-format ground cameras of their own. Soccer fans at the Corinthians Arena in São Paulo, Brazil, are watched over by an arsenal of wide-area cameras built by Hikvision, the Chinese firm that is fast becoming a leader in the field. Logipix, a Hungarian company, sells a panoramic camera with 200 megapixels. The startup Aqueti sells a gigapixel camera, the Mantis, based on the design of DARPA's 40-gigapixel AWARE system.

The technology is already trickling down the market. The 2017 Little League World Series in South Williamsport, Pennsylvania, was watched over by a network of 20-megapixel wide-area cameras developed by Canon out of many of the same components that can be found in professional photography cameras. Axis Communications, the company that sells the 20-megapixel camera, also sells a model that looks like any other CCTV system but can read the license plate on a car speeding through a roundabout from three football fields away. Some of these systems are affordable enough for use at home. Many of the largest CCTV manufacturers now produce wide-area cameras with relatively high pixel counts that operate like small-scale all-seeing eyes. Firms like Panasonic, Pelco, and Hikvision sell tiny high-megapixel 360-degree units that cost about $1,000.

The automated surveillance technologies discussed in the previous two chapters are also finding their way into ground-based cameras of both the wide-angle and soda-straw variety. Thanks to add-on facial-recognition and activity-detection software, the wide-area camera at the CenturyLink Stadium can tell the difference between someone placing a soda can and a large backpack in a garbage bin, or warn security officials if a known troublemaker is seen in the stands. The cameras at the Corinthians stadium are also equipped with facial-recognition features, as are many of Hikvision's low-end cameras, which come with automated analysis software that can detect intrusions, track individuals, and even detect traffic violations.

Meanwhile, firms specializing in computer vision are building systems that can be simply plugged into existing surveillance networks, a startling prospect given how much of modern public space is already watched over by CCTV. Interest in these products is intense. Following terrorist attacks like the 2017 bombing of an Ariana Grande concert in Manchester, UK, police manually comb through tens of thousands of hours of surveillance footage from hundreds of CCTV cameras in a desperate search for clues that might help to unravel the attacker's network and prevent an immediate follow-on attack. An automated system would reduce the time and manpower necessary for such a search by orders of magnitude.

Since 2016 Amazon has been selling facial- and object-recognition software to US law enforcement departments for processing imagery from CCTV feeds and other sources. According to Amazon's marketing materials, the software can discern a subject's emotional state based on their facial expression. The Orlando Police Department and the Washington County Sheriff's Office are already customers, as are a number of news services, including the *New York Times,* which use it to identify celebrities in videos and photos of large events.

Similar systems are being tested and installed across the globe in cities like Johannesburg and Singapore. In Moscow, officials have suggested that the technology could be used to track garbage collectors to make sure they are not slacking on the job.

With strong facial recognition, some of this software is capable of tracking a single target as he or she moves from one camera feed to the next. In cities where CCTV coverage is particularly dense — like London, which has collaborated with the startup SeeQuestor to test a CCTV tracking capability in its streets and Underground stations — such a program could continuously track a target for long periods, converting what was once a multitude of isolated cameras into a single, citywide all-seeing eye.

These systems will continue to grow smarter. IARPA, the US intelligence community's advanced-research laboratory, hopes that by 2020 it will have deep-learning software capable of recognizing 40 simple actions, 20 behaviors, and 12 complex activities in CCTV footage, including events such as "a person abandoning a backpack" and "people carrying or brandishing weapons," "a person throwing a rock or another object," "rioting and unrest," and even "subtle and anomalous behavior indicating out-of-context transport of unusually heavy objects."

As with their counterparts in the air, an intelligent wide-area ground camera could also have more benign applications. According to the team at Duke University that built the DARPA gigapixel cameras and now runs the firm Aqueti, one could eventually use gigapixel cameras to broadcast sports games, allowing viewers to zoom in on whatever interests them most in much the same way that ground troops can control individual "chip-out" streams from a Gorgon Stare feed.

The researchers contend that "zoomcast," as they call it, would be a much more engaging experience than a traditional broadcast. Maybe you are particularly interested in a certain defensive player, or perhaps you want to keep a close eye on the referee to make sure that he is making fair calls. With multiple gigapixel cameras pointing at a game, broadcasters could, eventually, create live, moving 3-D videos of the proceedings. This would permit viewers to navigate *through* the game as though they are actually on the field.

The Pentagon has already experimented with somewhat similar technologies for use on the battlefield, including software that turns aerial imagery of cities into detailed 3-D digital models that can be used to quickly familiarize soldiers with the environment they're about to step into. Steve Suddarth's software that turns WAMI video into three-dimensional moving cityscapes is likewise funded by the Pentagon. Police departments could use a similar zoomcast setup to navigate through a crowd in order to get a much closer view of potentially suspicious actors from a variety of angles.

The Duke team has already demonstrated a reduced version of zoomcast for the NBC series *Premier Boxing Champions*, using a large 360-degree camera installed a few feet above the ring. The cameras generate an instantaneous 3-D view of the fight. When one of the fighters lands a particularly heavy punch, you can freeze the frame, rotate around to get a better angle, and zoom in for the close-up. As journalist Matt Hartigan points out, it turns the boxing match into a real-life version of the bullet-dodging scene from the 1999 film *The Matrix*. The effect is as satisfying as it is unnerving.

WAMI Writ Large

While it takes root on the ground, wide-area surveillance is also fixing to break through the final frontier. "All that we've learned from WAMI in air," Sheila Vaidya told me, "can be transferred to cameras in space." In practice, a space-based all-seeing eye represents a technical challenge of celestial proportions. But it will, eventually, become a reality: one that everyone on earth — not just those unfortunate enough to live in war zones or crime-ridden cities — will have to contend with.

There has long been talk of putting WAMI into space orbit. Indeed, the original all-seeing eye program, at Lawrence Livermore, sought to build a surveillance satellite, not an airplane. Though all known WAMI systems produced to date have been mounted on aircraft, the community never fully abandoned the idea of the Big Daddy satellite as envisioned in *Enemy of the State*. While he was developing ARGUS at DARPA, Brian Leininger had extensive conversations with the National Reconnaissance Office — the branch of the intelligence community that manages its constellation of spy satellites — at a top-secret classification level.

Simply putting a camera on a satellite is fairly easy. In 2013 the SkySat-1, a satellite about the size of a dishwasher, produced incredible high-resolution video clips of several major cities. In the clip from Beijing, airplanes can be seen taking off and landing from an airport. In another clip, cars wind their way through a busy urban downtown.

But since satellites orbit Earth at 17,000 miles per hour, they can watch a target for only about 90 seconds before crossing behind the horizon, like a setting moon.* A number of groups are working, qui-

---

* While it is true that certain satellites can remain over a single location indefinitely, in what's known as geosynchronous orbit, they need to be at a much higher altitude. While it could technically be feasible to place a WAMI satellite in geo, with current camera technologies it wouldn't be able to pick up much detail on the Earth's surface from so far away. China's giant Gaofen 4 satellite, which currently sits in geosynchronous orbit staring at 49 million square kilometers of eastern Asia, is said to have some video-taking capabilities, though it probably can't pick out anything smaller than a very large ship.

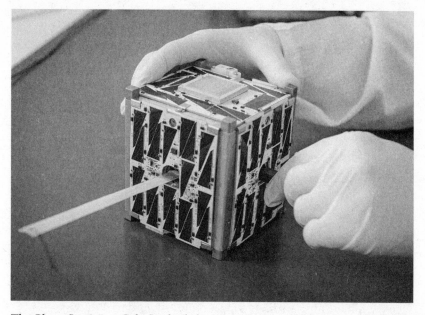

The PhoneSat 2.5, a CubeSat built by researchers at NASA's Ames Research Center in 2014. *NASA Ames*

etly, to develop techniques for extracting as much information as possible from such brief glimpses of a target, but even with one clip every 90 minutes — which is about how long it takes a satellite on a particularly tight orbit to make a full lap of Earth — you're still not going to be able to persistently track adversary individuals or groups.

A second problem is that the cosmos has always been an extremely expensive place to put a camera. A single imaging satellite built by Digital Globe, an established satellite-imaging firm, costs over half a billion dollars. Most US government satellites cost well over $1 billion.

Happily — or not so happily, depending on whom you ask — satellites are becoming much smaller and cheaper. CubeSats, modular satellites made up of 10-centimeter cubic blocks, cost as little as $40,000 to build and about $80,000 to put into orbit. A single launch vehicle can loft several dozen small satellites simultaneously. The contractor General Atomics has proposed using a giant electromagnetic rail gun to launch CubeSats into orbit, a technique that has the potential to drive down the launch cost even further.

As a result, in the coming years the number of tiny US spy satellites orbiting our planet will skyrocket. The Air Force, the Navy, the Army, and even the National Reconnaissance Office — which has been flying multibillion-dollar satellites since 1961 — all now maintain lively, and growing, small-satellite programs. Organizations that would formerly never have been able to field a satellite can, thanks to the "small-sat revolution," launch their own large constellations. Special Operations Command runs a fleet of satellites with a single job: tracking the tiny tagging devices planted on its fugitive targets.

Under a contingency plan revealed in 2017 called "Kill Chain," the Pentagon proposed launching a large constellation of tiny radar satellites that can generate images of key targets on the Korean Peninsula on an hourly basis. Achieving a similar level of coverage using large traditional radar satellites would have cost nearly $100 billion. Some in the Pentagon envision that eventually every analyst across the services will have at their fingertips a regularly updated database of every square inch of the planet's surface. DARPA is working to build constellations of dozens of small satellites that can provide troops on the ground with access to satellite imagery "with the press of a button" in 90 minutes or less, as well as satellites weighing around 600 pounds that can be launched with minimal advance notice in emergencies.

The interest in small satellites has not been confined to US shores. In 2017 a single Russian launch rocket carried 72 small satellites supplied by five different countries, while India's space agency is developing its own in-house launch system for small satellites. China's largest missile and rocket manufacturer, China Aerospace Science and Industry Corporation, predicts that it will be conducting 50 launches a year by 2020.

The commercial sphere is already light-years ahead of the government in this new space race. With its constellation of about 200 satellites, a Silicon Valley startup called Planet can photograph the globe's entire landmass at least once a day (weather permitting, of course: these satellites can't see through clouds or at night, though radar satellites like those being built by the startup Capella Space, which plans to use a constellation of synthetic-aperture radar satellites to provide "hourly" snapshots of particular locations, could do both). In 2016

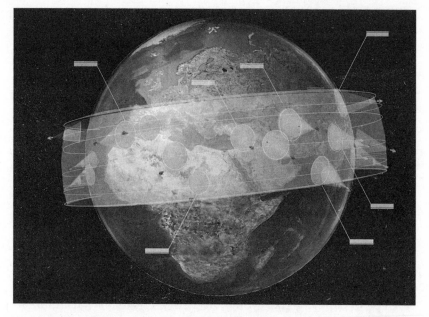

An artist's rendering of a proposed constellation of US military surveillance satellites that could be directed to almost any point on the planet in under two hours. *US Defense Advanced Research Projects Agency*

the National Geospatial-Intelligence Agency awarded Planet $20 million for ongoing access to its global imagery. Within a month of the program's launch, analysts had used the imagery to discover illegal gold mines in Peru and newly constructed airfields in Honduras.

Eventually, the technology and the rate of launches will have accelerated to a point where there will be enough satellites of sufficient power to generate one-meter-resolution imagery of the whole planet several times a week, and of every major city every single day. That means that every morning, we will have a new image of New York City, or Dubai, or Sydney, or Madrid, at sufficient resolution to tell who put their trash out and who didn't.

Once you have enough satellites to pass over a city daily, why not hourly? Another small-sat startup, BlackSky, will focus its constellation of 60 satellites on areas with the highest population density. The satellites would hit Baghdad 42 times a day, Pyongyang and Beijing 52 times a day, and Seattle 80 times a day. All told, 95 percent of the world's population will be photographed on a regular basis.

With a large enough number of satellites, a potential solution to the first problem of taking WAMI to space—the physics of a 17,000-mile-per-hour orbit—begins to emerge. Down within the atmosphere, when the Air Force wants to watch a single target for days on end with a drone, it uses multiple aircraft. As one drone runs low on fuel, a second drone comes in to take up the watch duty. With just four drones, the Air Force can watch a target perpetually. The same principle could be applied in space.

Charles Norton, an engineer who leads a team working on small-satellite technology at the NASA Jet Propulsion Laboratory, explained how it would work. One could arrange a constellation of small, inexpensive satellites in a straight line, "a string of pearls" formation in the oddly poetic lingo of orbital mechanics. The spacecraft are spaced in such a way so that every 90 seconds, as one satellite dips beyond the horizon, the next enters into view of the surveilled area, like a tightly choreographed orbital conga line. Stitching those 90-second clips together would, in theory, create a steady, uninterrupted view of a single spot on the earth. And thus, WAMI in space.

Other, less obvious obstacles to space-based WAMI are beginning to fall, too. One curious challenge is how to keep the satellite's camera pointing at a single spot on the surface of the Earth as it crosses overhead at 17,000 miles per hour, much the same way Nathan Crawford, the former golf tournament cameraman turned defense contractor, kept his camera locked on a gas station in the San Fernando Valley from thousands of feet in the air in an early test at Livermore. The task is tricky even for an experienced human, let alone for a small CubeSat with limited onboard space. But in 2013 the Aerospace Corporation, which has been developing small satellites for over two decades, kept a tiny one-kilogram satellite pointed at a single small peninsula on the southwestern tip of Australia as it sped by at full orbital velocity, demonstrating that "attitude control" (the actual technical term) is eminently possible.

Meanwhile, Lockheed Martin is developing a camera called SPIDER that consists of little more than a flat array of microchips. By eliminating the lens, which on a satellite needs to be large enough to focus on very distant objects, a SPIDER could be up to 100 times smaller and lighter—and, as a result, much cheaper to launch—than a regular camera of equivalent power. A single array capable of creating a high-resolution city-size aerial image would be about 12 inches wide and 1 inch thick.

What we may see even sooner than true-space WAMI is the high-altitude pseudosatellite (HAPS), a type of drone capable of flying at twice the altitude of a commercial airplane, right on the very edge of space, for days, weeks, or even months on end. It, too, is a difficult technology to master — Facebook and Google have both tried, and failed, to develop such systems — though not an impossible one. In one test flight, the Zephyr S, a HAPS developed by Airbus, flew for 25 days, 23 hours, and 57 minutes without refueling. (The UK Ministry of Defence already has three Zephyrs on order.) Such an aircraft, if equipped with WAMI, could provide a watching capability similar to that of a constellation of CubeSats with no need for a Russian rocket booster, a giant rail gun, or any "attitude control."

The tactical advantages of wide-area space surveillance, once attained, would be significant. Unlike the first generation of WAMI systems, which could not operate in defended sovereign airspace, Reaper drones and civilian airplanes being easy targets for antiaircraft missiles, a satellite can fly over anywhere on the planet with total impunity. Because satellites don't run on gas, a single constellation could watch an area for years. And since the satellites operate at such high altitudes, the surveilled views would be much wider and, as a result, even harder to evade.

Because of the massive scale of this data collected from orbit, truly astonishing feats will be possible when this space-based WAMI is parsed by the kinds of AI described in the previous chapters. Even satellite photographs can be frighteningly revealing when put through the right algorithms. Orbital Insight, a San Francisco–based startup, is able to regularly tally global oil reserves by running satellite imagery of the entire planet through deep-learning software that measures the level of every crude oil storage tank visible from space.

Using the same technology and the same imagery, the company can count every car in every department-store parking lot in America, predict a season's crop yields for entire continents, and monitor each of the hundreds of thousands of properties covered by a large US insurance firm. Naturally, Orbital provides services for a number of defense and intelligence agencies, though the firm declines to name them, or say what exactly they are looking at.

The CIA and NGA are working to develop similar AI programs in-house; to test and train them, they have assembled a database of images of hundreds of thousands of buildings around the globe — including much of Paris, Rio de Janeiro, and Khartoum. Meanwhile, the Defense Innovation Unit, a Pentagon startup incubator based in Silicon Valley, and the NGA launched, in 2018, an open competition for automated satellite analysis. The prize for the best algorithm: $100,000. Work is also already underway to do the same for satellite video. Researchers from Kitware are actively working on software for satellite video for the Air Force Research Laboratory using many of the same algorithms developed for tracking insurgents from within the atmosphere.

With automated WAMI from space, one would be able to detect anomalous behaviors and map networks of terrorists, criminals, or anybody else across entire countries. One could pick up the very earliest signs of a protest in Tehran, or of a battalion of tanks taking up an attack formation in the Baltics, or a forest fire in Australia, long before the news reached you by traditional channels. Companies like Orbital could mine the data to track all variety of economic and intelligence indicators — not on a weekly basis, but an hourly one. All this, and more, on any spot on the planet you want.

It's clear where this story is headed. Together, these kinds of developments, the engineer Bill Ross, who was involved in satellite surveillance programs at Lincoln Laboratory, told me in 2016, "start to get at wide-area motion imaging writ large, worldwide, or at least all the important cities." A system that would truly see everything.

At the time, I agreed with Ross in the way that I might agree with someone who tells me that someday everybody will get around in flying cars — they're probably right, but that *someday* could be a very distant one. Then, in the spring of 2018, a startup by the name of EarthNow announced that it would soon be launching a satellite constellation capable of generating "real-time, continuous video of almost anywhere on Earth" — that is to say, finally, just like the satellite in *Enemy of the State*. Initially, footage from the satellites, which are being built by Airbus, will only be available to governments and large companies. But eventually, EarthNow hopes to provide the service to a mass market, accessible, like Google Earth, to anybody — good guys and bad guys alike.

As of this writing, EarthNow has been evasive on the technical details of its scheme. Even though the venture is backed by a number of high-profile investors, including Bill Gates and SoftBank, there is good reason to be a little wary of its promises. But the story that the company hopes to advance to its final chapter is an inevitable one. Even if EarthNow doesn't succeed in watching the entire globe, perpetually and unblinkingly, we know how this all ends: sooner or later, someone else will.

10

Watching in Every Sense

CHEFS LIKE TO say that you need all five senses to cook well. When you panfry a steak, the sound of its sizzle is the best indicator that your skillet is hot enough, and a reliable way to tell when it's done is to poke it with your finger. If something starts to burn, your nose will likely tell you well before your eyes.

Similarly, the US government has at its disposal a variety of different ways of sensing the world. Among the Air Force's 500-plus intelligence, surveillance, and reconnaissance aircraft are radar systems that can detect all moving objects across an area the size of Slovakia, hyperspectral sensors that see electromagnetic radiation emitted by hundreds of different types of materials, and radio-frequency interceptors that can key you into your adversary's radio chatter on the ground from six miles overhead.

The story is no different in the intelligence agencies. The NGA owns laser imaging sensors that can generate 3-D maps so detailed they will even detect certain invisible aerosols floating through the air. The NSA's Gilgamesh airborne cell-phone interceptor can pinpoint the location of a mobile to within a couple of meters, while another of its aerial systems, Shenanigans, can reportedly absorb the data from every WiFi router, internet-connected computer, and smartphone in a medium-size town in one fell swoop.

At the highest classification levels, some of the tools are truly sci-fiesque, bizarre even. In an unusually candid briefing in 2007, an official at the Joint Special Operations Command, the Pentagon division that oversees the military's black operations, noted that the organization would soon be fielding self-deploying listening devices, biomimetic drones, and even synthetic dog noses that can pick up the scent of a human from miles away.

This is all in addition to social media, arguably the greatest source of intelligence the world has ever seen — and a self-seeded one, at that. A single Facebook profile may include a target's detailed biographical information, large collections of images showing what the individual looks like, updates on their evolving political views, logs of where they spend their time, along with a roster of family members, friends, and associates with profiles equally rich in data. It is a source that the spy community monitors in fervid and increasingly sophisticated ways.

Take just this one example: a DARPA effort by the unsubtle name of Worldwide Intelligence Surveillance and Reconnaissance, which went black around 2014, set out to turn millions of YouTube and Instagram videos into 3-D reconstructions of places and events where the United States has no access to the airspace — think countries with air defenses, like China, Russia, Iran.

The US government is at its most inescapable when it uses all of these technologies and sources of information together. While WAMI excelled at hunting and tracking insurgents in Iraq, Afghanistan, and elsewhere, the Pentagon has always thought of it as just one part of a much larger surveillance apparatus, an approach referred to as "layered sensing." In a single mission, the Air Force intelligence unit responsible for processing Gorgon Stare imagery may also use data from a range of different types of radar, photographic cameras, cell phone interceptors, and spectral thermographic chemical detectors.

This diversity of eyes in the sky enables a truly unblinking view of the ground. Using WAMI technology, when you're following a car, a cloud might block your view every so often. But radar can see in all weather conditions. If you can't quite make out your subject in detail in the Gorgon Stare footage or via radar, the soda-straw camera on a nearby Reaper is powerful enough to show operators not only what kind of hairstyle the subject has, but also whether he or she is holding a cigarette. A cell phone interceptor will reveal only the general location of a suspicious device; if you want to see the specific vehicle in which the phone is detected, you have to revert back to WAMI.

Combining data from different sensors is at the heart of how the Pentagon finds targets in the first place. In a process known as "cross-cueing," a broad-area sensor like a WAMI or a ground-tracking radar will scan the entire battle space for suspicious activities, and then hand the coordinates of any unusual actors to an aircraft with a soda-straw camera, which could confirm whether the driver in question is an insurgent or merely a civilian who got lost when his cell phone ran out of batteries.

Customs and Border Protection uses a similar technique over US borders. Whenever one of the agency's Arizona-based Reapers detects something moving with its VADER ground radar, which can monitor vast stretches of land, the crew will slew a telescopic camera over to the given coordinates of the action to identify those involved.

Using multiple sensors in this way is particularly important when the findings are made by algorithms. If a single camera with a less than perfect algorithm identifies a particular blob on the screen as

Analysts from the 362nd Expeditionary Reconnaissance Squadron managing multiple intelligence feeds at a deployed intelligence cell at Camp Liberty, a military installation north of Baghdad, in 2009. *The appearance of US Department of Defense (DOD) visual information does not imply or constitute DOD endorsement.* *US Air Force/Senior Airman Jacqueline Romero*

an enemy tank, you can't rule out the possibility that the suspected tank may actually be a refrigerated truck. But if a second sensor also detects a tank, then you have much more reason to believe that you are, in fact, dealing with a tank. The automated behavior-detection developed by Signal Innovation Group that was tested on Lubbock, Texas, has the best results when fed with multiple surveillance feeds in parallel.

The process of meshing all of this information is difficult, tedious, time-consuming, and expensive. Each different type of intelligence requires different analysts with different expertise, sometimes from different agencies. In a large, complex operation, you may need a CIA interpreter for the radio chatter, an NSA cyber expert for the WiFi data, and an Air Force imagery specialist for the video footage. Even though they are all looking at the same set of targets, each analyst will produce a complicated data set that might not make any sense to the others.

Such an arrangement is like a team of chefs who each possess only a single sense: one chef is limited to smelling the food, another to tasting it, a third to touching it, and a fourth to seeing it. In time-sensitive military operations, teams working on distinct sources of intelligence for a single target often have to literally yell at each other across the floor of the intelligence center — or, if they're not in the same room, yell at each other over the chat boards — to share their findings. If one team loses sight of its target because, say, the target has parked under a cloud, analysts must rush to turn the track over to a second sensor capable of seeing through that cloud. If they're not quick enough, the target might launch an attack, or disappear from view entirely, becoming untraceable by any of the teams' sensors.

A core tenet of the Pentagon's intelligence strategy for the 21st century is to offload the melding of this disparate intelligence to computers.

The idea, known as automated sensor fusion, is that a single piece of software — fueled by AI — would collate data from a range of different sensor types to produce a single intricate portrait of the target. Instead of multiple agencies staffed by thousands of employees, each staring at distinct types of pixels, trying desperately to build a common picture out of the chaos, you have a single app, accessible from anywhere, that does all of that for you.

Given how many types of intelligence-collection tools now exist in the Pentagon and intelligence community's arsenal—not to mention, increasingly, among law enforcement agencies at home—and given how widely they are used in every corner of both the physical and digital worlds, the payoff of this technology, if it works, would be massive. Groups that previously managed to slip away unnoticed may not get so lucky anymore. And, hopefully, those who would have been targeted accidentally under the traditional manual method of analysis will live to see another day.

Letting the Computers Take Over

Take Hydra, a software system built by BAE. Hydra uses Bayesian logic to automatically correlate information from all of the types of sensors that the Air Force processes alongside WAMI in its Gorgon Stare analysis cells, as well as half a dozen other sources of intelligence.

If a signals-intelligence sensor detects a suspicious call but isn't able to pin it to a specific vehicle, the software will check the data against a video feed and employ probabilistic reasoning to highlight the likely target. Once the correct vehicle is located, the program will then use the same correlation techniques to ensure that it never loses sight of the party of interest. As a target slips out of view in one sensor, the software instantaneously turns to a different sensor to pick up where the first system left off.

Under the same Air Force program that funded Hydra, researchers at Sandia National Laboratories have built software that cross-checks the identity of a target across multiple spectra in the lead-up to a strike. Theoretically, the system would help ensure that when the Air Force pulls a trigger, it's aiming at the right person, rather than, say, the intended target's grandmother, who, on the basis of cell phone intelligence alone, might match the profile of a hardened combatant.

Some efforts are focused on fusing various sources of intelligence right onboard an aircraft equipped with a multitude of sensors. Bright Eyes, a system under development at Logos for an undisclosed intelligence agency, incorporates day and night wide-area cameras; a hyperspectral sensor, which detects chemical emissions; a Light Detection and Ranging unit that generates precise 3-D models of the target; a telescopic camera; and two classified sensors.

The aircraft will automatically take close-up images of every target tracked in the wide-area footage, while also determining whether those targets are transporting explosives and (probably) intercepting their electronic communications. Rather than beam down multiple separate surveillance feeds for analysts on the ground to fuse manually, such systems produce a single multifaceted view of any targets they are watching. The British Ministry of Defence is building a multisensor drone modeled on the same concept designed for what it describes as "fully autonomous target detection." Appropriately, they call it Project Omniscient.

These programs can condense giant volumes of surveillance into interfaces that are simple and intuitive. Working with Carnegie Mellon and several private firms, the Air Force Research Laboratory is developing a program called Tentacle, which autonomously identifies and tracks vehicles and individuals across hundreds of video feeds, then combines these tracks into an immersive 3-D model of the area. Instead of having to monitor dozens of videos at once, the analyst would have a single commanding view of the ground, explorable from any angle. In unclassified simulations, it looks just like a video game, simple enough for even a child to understand.

In traditional intelligence regimes, there is a constant risk that, with so many different types of information segregated among so many different agencies and disciplines — "stovepipes," in spy speak — important pieces of the puzzle may get lost in the scrum. Two analysts in two different places staring at two different intelligence feeds might have information that, if combined, would point to an as-yet-unknown threat, but never have the chance to share their intel. This lack of coordination among disparate agencies has been the root of many of the major intelligence failures of recent years.

Insight, a DARPA program focusing on this very issue, will autonomously collate intelligence from wide-area imagery, electronic communications, radar data, soda-straw video from drones, and human sources so that no two tidbits about a single target ever float in isolation. Designed to be "rich" and "interactive," like a user-friendly smartphone app, the program will generate something resembling a Facebook profile on every target (software like Insight will also, obviously, draw heavily from social media data). When these collated crumbs of intelligence point to a possible anomaly or threat — say, a visit to a known safe house, as seen through Gorgon Stare, followed by a rumor, picked up by a local informant, about an attack on a nearby market — the software will put two and two together and alert its human operators.

Among other ambitious promises, DARPA insists that Insight will be able to build pattern-of-life models that account for what DARPA calls social and cultural "dynamics" that vary from region to region, and even predict where a given target will travel based on their previous routes and communications. In the same vein, the NGA has software that tracks what people are saying on Facebook and Twitter in the areas being watched with systems like Gorgon Stare, putting the airborne imagery in the context of information that could be useful for analyzing factors like a city's "stability dynamic" — the intelligence community's term for the likelihood that mass violence might erupt at any given time.

Building on similar principles, an Air Force initiative seeks to build a deep-learning system so smart that it can fuse data to find "targets of interest" and suspicious activities under circumstances that would flummox a human analyst, including "deceptive conditions, electronic warfare," and environmental conditions such as bad weather that degrade the quality of the information itself.

As the Air Force envisions it, the software won't even need to be told what to look for. Instead, it will simply read the chat-room communications between intelligence analysts and infer the types of information they might find of interest. If two analysts are chatting about a particular individual in the morning, the program might gather intelligence on that person from a separate agency and have it delivered to them by lunchtime.

All Together Now

What happens when all of this work comes together? In an article published in 2018 called "An Orchestra of Machine Intelligence," a team of engineers from BAE, two satellite imagery-analysis firms, and the Kostas Research Institute for Homeland Security at Northwestern University provide one possible vision of this unified, futuristic approach to surveillance.

In the authors' scenario, an intelligence analyst sets out to locate Osama bin Laden. He opens his artificially intelligent fusion system and asks, "Where is Osama bin Laden?" as though it were Siri. The software begins sorting through vast quantities of available intelligence information to identify the sources most likely to yield tips relevant to the query—again like Siri, which often performs a Google search to answer your questions. After determining that bin Laden is probably somewhere in Pakistan or Afghanistan, the software runs a voice-recognition program on all recorded cell-phone conversations from the area and discovers a number of calls between some of bin Laden's associates. The software identifies one call, from a public phone booth, as being of particular interest.

In order to identify the unknown caller at the other end of the line, it cross-checks footage of the area from a hacked web camera as well as an all-seeing eye flying overhead. Motion-tracking software then follows the caller both forward and backward in time through the footage. The caller stops at a mysterious residential compound, which a little googling reveals to be owned by a senior military officer whose recent social media posts indicate "fervent" support for bin Laden. The software returns to the human analyst with the compound's address and a score reflecting its confidence that it has the correct answer to the original question.*

Though it will be a few years before the Pentagon can pair analysts with digital assistants that can autonomously find the most wanted men on earth upon command, some automated sensor-fusion weapons are already making their way onto the battlefield.

Advanced fighter jets like the F-35 Lightning II and the newest F/A-18 Super Hornet, for example, come equipped with software that blends data from the aircraft's various different sensors into a

---

* At the time of the actual raid against bin Laden, the CIA, which had manually found the compound through a similar process, also had a confidence rating for its analysis: 60 to 80 percent.

single view of its surroundings. In 2017, after at least half a decade of development, the Air Force appeared to begin installing the BAE and Sandia software, along with similar programs for air-to-air targets, on its Reaper drone control systems. Meanwhile, a tool developed under DARPA's Wide Area Network Detection program that extended the length of vehicle and dismount tracks in WAMI data by fusing them with signals intelligence was transferred to Special Operations Command in 2013. A source confirmed that it is currently in use in the field for imagery captured by an ARGUS camera.

Among the intelligence agencies, a number of these tools have already been in use for years. NGA's Map of the World, which looks a bit like Google Earth, compiles intelligence from a variety of sources for every inch of the planet's surface, allowing users to access troves of imagery for any location of interest, both within and beyond areas of active hostilities. The agency's Multi-INT Analysis and Archive System and another program called QuellFire, which is owned by an undisclosed intelligence organization, automatically draw on a broad set of databases to assemble Facebook-like case files on known terror suspects. By 2020 the NGA predicts that any analyst in the agency will have at their fingertips almost every data type produced by the intelligence community, all in one place, like some kind of all-seeing superspy.

Other fusion products are already on sale in the open market. BAE sells a software suite called Movement Intelligence that's similar to the DARPA-developed Wide Area Network Detection system, as well as a commercial version of the Hydra software. (BAE, which also built ARGUS, is an industry leader in this space. It was the prime contractor for both the Map of the World and DARPA's Insight. As of 2018, the company had received over $400 million in contracts for work on the two programs.)

SRC, a New York–based research corporation, sells software that blends drone footage into Google Earth, along with signals intelligence from sources such as phones and radios. The Virginia-based defense firm Leidos offers a product called Advanced Analytics Suite, which is capable of automatically fusing visual and textual intelligence; the company boasts that it is particularly well-suited for discovering high-value targets. The defense giant Lockheed Martin, meanwhile, sells a product that pulls together information from satellites, drones, and human analysts to create a giant, immersive 3-D simulation of the battlefield. In a brochure, the company notes that because the program was developed in Canada, it can be exported to pretty much any country on the globe.

This software does come at a price. A license for BAE's Hydra costs $144,000, while the company's tracking software for WAMI footage is $287,000, plus $49,000 for one year of maintenance. Lockheed Martin's software, at just $9,000, seems like a bargain in comparison.

It Takes a Swarm

Once computers become smart enough to work alongside humans as collaborators, it's only a small leap to have them team up with other computers.

Imagine that in some not-too-distant future a team of drones is dispatched into adversary territory in a surprise attack against a series of antiaircraft radar sites, a mission that may be too dangerous for regular piloted airplanes. In order for the drones to evade detection, there can be minimal communications between them and the operators at home base. The drones are on their own.

In the hypothetical antiradar mission, which comes from a simulation published by DARPA, the drones behave much like a team of Air Force analysts chasing a target in 2018. One drone is equipped with a wide-area sensor. When it detects a possible radar site, it cues a second drone with a soda straw to take a closer look. Because the weather is bad, the soda-straw image isn't particularly clear, so a third drone scans the site with a synthetic-aperture radar, which, when added to the video footage, sharpens the image enough for the swarm's target-recognition algorithms to confirm the target, which they then destroy. DARPA notes that the drones will strictly follow the DOD's rules of engagement.

Each drone understands the importance of the data it's collecting and knows when and how to share it with its teammates. When one device has trouble identifying a particular object, it will confer with the other drones and robots in the group, just like an Air Force imagery analyst might turn to a colleague for a second opinion as to whether the pixels on their screen represent a motorbike or a donkey. If another drone detects an antiaircraft missile launcher in the area, it will convey that information to the others, and they will reroute to remain out of its range.

Such a system would keep human pilots out of harm's way and cut down on the need for human intelligence analysts. It could also speed up the pace of battlefield decisions by taking the sometimes frail, wavering link that is the human mind out of the equation. And because drone swarms of this kind don't need to have a constant communications link to the home base, they could be particularly useful for conducting autonomous covert missions behind enemy lines.

A second, more nightmarish DARPA swarming effort called OFF-SET, for OFFensive Swarm-Enabled Tactics, is focused on building similar teams optimized for urban battle spaces. In one notional mission proposed by the agency, a swarm of more than 100 drones instructed to "assault and clear" a building will collaborate to find the optimal entry points to the structure, locate its occupants, build a map of the space, and pinpoint the source of radio emissions and chemicals. All told, the mission would be over in less than two hours.

At the very least, software like this would make a swarm of drones very difficult to outrun. A program called Bedlam is a case in point. Developed with funding from the Army, it is designed to optimize swarms for chasing targets on the ground. In a publicly released simulation, a single drone equipped with the software pursues a vehicle through the winding streets of what looks like a Middle Eastern city.

The video is worth watching. Each time the vehicle makes a turn, the drone instantly changes its course and calculates the target's new possible escape routes. If a swarm were equipped with this type of reasoning capacity, when one drone detects that the vehicle has changed its course, it could communicate directly with the other drones to get ahead and cut it off, like a pack of wolves outflanking a caribou.

These programs become even more inescapable when the swarms incorporate surveillance systems on the ground. In 2017 researchers at Cornell announced a plan to conduct live trials of an automated fusion system using teams of surveillance drones, robots, and ground cameras on the university's campus in Ithaca, New York. Using deeplearning algorithms and Bayesian logic, a drone that spots what looks like a group of people fighting might turn to a ground robot with a clearer view of the action. A ground camera that notices somebody leaving a backpack in a crowded square will pass the tip over to a drone that can track the "suspect" away from the scene. There is already talk of incorporating certain features of the Cornell product, which is being paid for by the Navy, into the university's campus security system.

Why Does Fusion Matter?

The world today is full of sensors. On a short walk through any modern city or university campus I might pass dozens or even hundreds of cameras; the cell phone in my pocket beams out my location constantly. License-plate readers scan my car's tags without me even knowing it. Occasionally I volunteer my political views on Twitter. I probably post a selfie online just regularly enough to keep the facial-recognition software up to scratch.

For the most part, these sensors don't currently talk to one another. When I walk past 50 cameras, they aren't smart enough to piece together that the individual crossing each of their frames is the same person. They don't know my Twitter handle, or the license number of the last car I drove. Although Taekwon Hart's assailants undoubtedly have shown up on thousands of CCTV screens in the years since his shooting in Brooklyn, none of those cameras can connect them to the crime. The men are anonymous pixelated ghosts — unknown unknowns.

Sure enough, the principles of automated sensor fusion are already catching on in the domestic sphere. About 80 local law enforcement agencies in 49 states, the District of Columbia, Puerto Rico, the US Virgin Islands, and Guam operate so-called fusion centers modeled on military war rooms that serve as a clearinghouse for information collected by a range of different agencies and private organizations.

According to guidelines drafted by the Department of Homeland Security, these centers, which provide local agencies with access to "national threat information," are geared toward fighting "violent extremism."

Meanwhile, a number of US law enforcement departments, including the Los Angeles Police Department, the Metropolitan Police Department in Washington, DC, and the Virginia State Police operate fusion software developed by the Silicon Valley firm Palantir that collates and analyzes criminal records, gang-member databases, number-plate rosters, social media archives, and even jailhouse telephone records to pinpoint individuals who are at risk of committing future violent crimes. (The company has worked extensively with the Pentagon and the intelligence community, and is rumored to have provided data-analysis services that assisted in tracking down Osama bin Laden.)

Meanwhile, New York, Singapore, and a number of undisclosed cities operate an elaborate fusion system built by Microsoft called Aware that links both imagery and textual intelligence drawn from both municipal data repositories and live surveillance systems. A representative I spoke with in 2017 said that the company was working on plans to bring Aware to various new markets. When I asked which markets, specifically, he declined to go into more detail either on or off the record and abruptly ended the conver-sation.

You could imagine how such a system might have helped investigators looking into the shooting of Taekwon Hart. A gunshot detector could have directed the CCTV cameras dotted around the neighborhood to begin tracking any suspects departing the corner of Lewis and Van Buren. The cameras would, in turn, run the images of all passersby against a mugshot database to see if they have a criminal record. By the time the police arrive on the scene, the system could know exactly where the assailants are hiding, and who they are.

In the months leading up to the 2022 World Cup, Qatar's security agencies plan to use a fusion system called ARMED that will comb through social media posts, text messages, and intelligence reports on known terrorist networks, searching for people who might be likely to commit an attack. When a prototype of ARMED was run against a trove of data from the days leading up to the 2013 Boston Marathon, it placed the Tsarnaev brothers, perpetrators of the bombing that killed three people and wounded more than 260 others, in the top 100 individuals deemed most likely to commit an atrocity. If, during the World Cup, the software establishes that a particular suspect is indeed planning an imminent attack, it will fuse data from aerial drones, ground cameras, and cell phone trackers to lead police to the individual's location before any damage can be done.

In China, a company called Dahua Technology sells a system that can recognize your face in a CCTV video and then automatically trace all of your movements across all other connected cameras up to a week back in time. It is even smart enough to figure out what car you own. The company has claimed that in cities with particularly dense CCTV coverage, its software will be able to identify those citizens, or dissidents, whom a target interacts with most frequently. Those people, in turn, can be tracked and matched with their associates. And so on, ad Orwellian infinitum.

In a fully fused city, there may be nowhere to hide. An aerial camera may autonomously detect my U-turn and then track me to wherever I park. A license plate reader will check if I have outstanding violations in another state. Once I am on foot, the ground cameras pick up the slack. If I upload a photograph to Instagram from a street corner, my identity and the arrangement of pixels representing me in the imagery can be correlated, and my whole social media past will be laid bare to whomever considers me a subject of interest, or suspicion.

It doesn't help that we continue to willingly produce and upload volumes of fusible data from the most intimate corners of our lives. Just consider the internet of things (IoT), the networks of personal devices, vehicles, home appliances, and other previously inert electronic tools that are able to send and receive data.

Though these systems might add convenience to our lives, they will paint an incredibly detailed portrait of every one of us. By some estimates, there will be 80 billion smart devices in operation by 2025, with 150,000 new devices coming online every minute. The Pentagon already has designs to take advantage of all this free surveillance. "This immense, sparsely populated space of interconnected devices," explains one Pentagon document, "could serve as a globe-spanning, multi-sensing surveillance system." The document quotes legal expert Julia Powles, who wrote in 2015 that the IoT will be "the greatest mass surveillance infrastructure ever conceived."

We therefore need to think about all-seeing surveillance in the context of all of the other technologies that surround us. Wide-area surveillance is significant not simply because it can record our every move or track thousands of cars at once. It matters because it is emerging at a time when much of our lives is already subject, in one form or another, to the unblinking gaze of a sensor, whether it's a CCTV camera, an aerial-surveillance plane, a satellite constellation, an email surveillance tool, or our own personal Instagram feeds.

WAMI is arguably the "Big Daddy" of all of these systems, but rather than any single device, it is the fusion of these technologies that will carry us inexorably toward total surveillance. That is, unless we take action.

PART III

Another Long Road

11

One Hell of a Fight

FOR AS LONG as wide-area surveillance has been publicly known, there are those who have tried to stop its encroachment into civilian life—or at least slow it down. Among them, few have worked against the all-seeing eye as unswervingly and forcefully as Jay Stanley, a senior policy analyst at the ACLU, who has been a vocal critic of the technology since 2013.

That year, Stanley discovered WAMI while watching an episode of the PBS documentary series *NOVA* that featured a short segment on ARGUS. In the segment, the lead engineer for ARGUS, Yiannis Antoniades, his face half-shrouded in darkness, describes some of the system's specifications while carefully leaving others out. "It is important for the public to know that some of the capabilities exist," he says. In the background the camera sits hidden under a blue tarp.

These capabilities, Stanley knew, had ominous implications. Eighteen months earlier, a team of photographers in Vancouver had created a 2-billion-pixel photograph of a large crowd of hockey fans by stitching together 216 individual images taken from a building overlooking the event. In the image, which is still available online, the faces of crowd members are more than detailed enough to feed into a facial-recognition system. A riot had broken out among the fans later that day, and Stanley, along with a number of his peers, noted that such a high-resolution image could be used to identify individuals who might have been involved in the violence.

Stanley realized that ARGUS represented the marriage of high-resolution surveillance and drone technology, which he had also been following, and opposing, for years. "Even in our most pessimistic moments, I don't think we thought that every street, empty lot, garden, and field would be subject to video monitoring anytime soon," he wrote in a blog post after watching the *NOVA* documentary. "But that is precisely what this technology could enable."

Stanley started tracking developments in the field closely. "We have a technology now to create the nightmare scenario of surveillance, where in the sky over every city they're tracking everybody's movements," he told me in 2017. "I was thunderstruck at the choices that this technology would be presenting us as a society down the road as it played out, as it probably would."

It wasn't long until Stanley had occasion to witness, and influence, one such choice. The summer before he discovered WAMI, Persistent Surveillance Systems had carried out a one-week operation over downtown Dayton, Ohio, and had submitted to the Dayton City Commission a proposal to conduct 120 hours of surveillance in the summer of 2013 under a program that PSS founder Ross McNutt and his team called "Trusted Situational Awareness."

When the proposed program was announced in January 2013, a group of local residents, with assistance from the ACLU of Ohio, pushed back. After meeting with the group, the Dayton Police Department issued a revised policy for the surveillance operation with stricter privacy protections, but the ACLU spurned the revision, as it did not include a warrant requirement or a detailed provision on data retention and sharing.

"In America," Stanley wrote in a post on his blog, "we do not allow the government to look over everybody's shoulders (literally or figuratively) just in case they engage in wrongdoing." One anonymous reader commented, "I live in Dayton and I find this absolutely despicable and Orwellian. Somebody needs to get fired. This is evil, pure and simple."

Two weeks later, the Dayton City Commission abandoned McNutt's proposal. "While we believe there are real potential benefits to the strategic application of this technology," the city manager said in his announcement, "we heard enough confusion over how it would be applied to concern us." According to materials that McNutt showed me in Baltimore, the commission's rationale was that the policy needed "further clarification and refinement." But the commissioner was clear that the city was unlikely to revisit the proposal anytime soon: "With the reaction we got, it's clear the citizens aren't ready for it."

McNutt was, and remains, furious. He felt that the whole program had been derailed by "twelve protesters" who didn't understand how WAMI works. "It's a little hard to argue with people who say, 'Drones kill people, you fly drones, therefore you kill people.'" But he was also undeterred. According to one of the people who participated in the council meetings, McNutt said that if Dayton was going to block the program, "he was just going to go to another city."

A few months later, at a conference in Orlando, Florida, McNutt approached Stanley — whom he noticed was wearing an ACLU badge — and introduced himself. He pressed his point that the technology posed far less of a threat than Stanley claimed, but the discussion was amicable, and the two men agreed to remain in touch. In early 2014 McNutt and Stanley met again at the ACLU's offices in DC. McNutt explained that the company worked within a set of strict self-imposed boundaries. Political dissidents and vulnerable groups were protected under company policy. The camera, he said, couldn't recognize faces or read license plates.

Stanley was unconvinced. As technology improved, he wrote in a post about the meeting, those limitations would be overcome, and even McNutt admitted that just because he didn't choose to abuse the technology didn't mean that someone else wouldn't: eventually "someone is going to do this," he reportedly told Stanley.

For the advocacy community in which Stanley operates, the dangers of wide-area surveillance are both manifold and obvious. When I asked Stanley what specific dangers wide-area surveillance technology posed, he seemed a bit nettled by the question. "You're asking, 'Why do we care about privacy?'"

"There is a long and unfortunate history in this country of law enforcement using surveillance powers and tools for political purposes," he continued. "Not necessarily where there is a good reason to believe that somebody is planning to break the law, but where law enforcement officials disagree with the political substance of people's views."

On this point, Stanley has since been joined by a growing chorus of academics and advocates from across the political spectrum who share his position. Figures such as Jake Laperruque, a privacy expert and senior counsel at the Constitution Project, and Matthew Feeney, a policy analyst at the Cato Institute, the libertarian think tank, have argued that the technology could be used against political dissenters and minority groups just as effectively as it is currently used against criminals. "Persistent surveillance allows users to track mosque congregations, protesters, abortion clinic visitors, Alcoholics Anonymous members, and gun show attendees," wrote Feeney in 2016. If ARGUS were ever used by police, Laperruque declared the following year, "it will mark the end of anonymity."

Others have focused their critiques on the lack of transparency that attends most WAMI operations in the United States. "It's especially galling," wrote the *Atlantic*'s Conor Friedersdorf following the revelation that Los Angeles had conducted a secret wide-area-surveillance test program over Compton, "to see law enforcement professionals betray the spirit of democracy by foisting these tools on what they know to be a reluctant public because they deem it to be prudent based on a perspective that is obviously biased." (The commander who oversaw the program was unsympathetic to these concerns: "With the amount of technology out in today's age, with cameras in ATMs, at every 7/11, at every supermarket, pretty much every light pole, all the license plate cameras, the red-light cameras, people have just gotten used to being watched.")

For Stanley, the greatest danger of wide-area surveillance is not the potential for isolated instances of abuse. It's that the all-seeing eye sees all; to use the privacy community's term of art, it's a dragnet. He concedes that WAMI could help solve certain crimes. But, he asked, "do we want to allow the government to record everything we do just in case it happens to be useful later on down the road?" By watching everything, the cameras will transform the very meaning of public space, to say nothing of the sacrosanct relationship between the citizen and the democratic state.

He went on: "It's just too much power to give to the government, because the ability to press 'rewind' on somebody's life is probably the ability to ruin them, in many, many cases. People say, 'Well, I've got nothing to hide.' But some people do, and we don't want to live in a society where activists who are challenging authorities have to worry about such a system, and even if you think you don't have anything to hide, you might be the victim of a mistake, and a determined prosecutor per chance can find something that you've done wrong, because the laws are so complicated and discretion in applying them is so wide."

Stanley's position on surveillance technologies, old and new, is based on the Fair Information Practice Principles, a set of five rules for electronic-information collection and use developed by the Federal Trade Commission. "There shouldn't be information collection going on about which you are unaware," Stanley explained, paraphrasing the principles. "It shouldn't be going on without your permission. Information shouldn't be used for purposes other than the ones that you gave permission for them to be used for. Information should be kept as securely as possible, and people should be able to access the information that's held about them."

As far as he's concerned, the all-seeing eye could violate all of these precepts at once, and he's determined not to let that happen. "I expect that we're going to have a hell of a fight over it," he said.

The Creators' View

The view from the other side of the field is, unsurprisingly, almost exactly the opposite. Everyone I've met who has been involved, even peripherally, with the all-seeing eye believes that they have created a force for good. Many seem to see their work — from their early efforts combating terrorist networks in Iraq and Afghanistan to their more recent campaigns to bring the technology back home — as a sort of righteous quest.

These figures approach their work with considerable zeal. While running the Community Support Program in Baltimore, McNutt and his wide-area-surveillance team investigated sexual assaults, burglaries, and traffic accidents even when nobody in the police department had instructed him to do so. Nathan Crawford, the founder of CRI, says that he will hang up his hat in what he calls "his journey" the day he is able to stop a child abduction in progress. One Persistent Surveillance Systems document claims that the company had already, as of 2012, been involved in kidnapping investigations for various private clients.

Yet the creators of WAMI also know that the line between good and evil is a blurry one. Initially, when the groups focused on IEDs, there were few ethical questions. "But it wasn't a big leap for people to think, Well, wait a minute," Bill Ross, the former MIT Lincoln Laboratory engineer, said, "what happens when these systems are turned on inside the US?" The question wasn't just, "How do we plan to use this?" Ross explained, but also, "How will others use it when we're done?" After all, the technology was inspired by a movie about the dangers of an overly powerful surveillance state. How could they be sure that the nightmare scenario of *Enemy of the State* wouldn't become a reality?

Crawford said that during the brief collaboration between the two earliest WAMI programs, the engineers on the beach all agreed that if they were going to put this technology out in the world, they would have to do it, he recalled, "the right way." In a later conversation, he added, "if it's not truly a capability that serves the interests of society, then it shouldn't exist."

He paused. "If it gets turned into that, I'll be one of the first people to speak out against it."

Until now, most of the engineers and officials responsible for the creation and development of wide-area surveillance technology have not spoken publicly about the moral consequences of their work. One gets the sense that, after so many years of silence, they are itching to get something off their chests. With few exceptions, they raised the issue of ethics in our interviews before I did.

In one conversation, with Michael Meermans, the former congressional staffer and current Sierra Nevada executive, I asked about the civil applications of a camera like Gorgon Stare. After listing a few examples, he stopped. "Now, there will always — *there will always* — be the issue of privacy rights." When I asked the ARGUS engineer Yiannis Antoniades, who was born in Greece and emigrated to America as a college student, whether he had thought much about privacy when he was working on the program, he explained, "We all love our freedom. That's why we're in this country. So, 'yes' is the answer."

When Crawford gives demonstrations to potential customers or sponsors, people often ask him to zoom in on their own neighborhood or even their home, and he always complies. In many cases, the individual who has made the request instinctively starts following cars and figures on the screen. "And all of a sudden they get this immediate feeling in their soul," Crawford said, "'I don't want to do this.'"

"Good," Crawford will tell them. "You just learned the first lesson of wide-area."

Not everyone who operates WAMI will have someone like Crawford at their side, and that, the developers I spoke to concede, is part of the problem. In 2016 William Ray Walters, an amateur musician and vlogger, learned from a friend that his wife had been meeting with another man before work. And so, in one of the first known instances of what is likely to become a commonplace occurrence, Walters used his personal quadcopter drone to follow and videotape his wife, without her knowledge, as she walked to work. As long as Walters wasn't breaking any airspace regulations, this was completely legal.

Walters posted the video to YouTube, where by late 2018 it had been viewed some 15 million times. Sure enough, Walters's wife, whom he later identifies as Donna, approaches an SUV in a pharmacy parking lot. Walters, in a shaky, sometimes teary narration, declares, "There's the guy that my wife is cheating on me with." Walter's wife walks up to the vehicle and gives the driver a kiss. She then gets in and the car drives off. "There it goes. Boom," says Walters, now screaming. "Eighteen years and you just throw it away like that."

In an interview with *Inside Edition* shortly after the video was posted, Walters said that he was "one hundred percent sure that I was going to kill that man."

One year after posting the video, Walters announced that he and Donna were back together, and all, it seems, was forgiven. In a video responding to questions from subscribers, Walters said that he regretted posting the surveillance video online. "Yeah, I probably shouldn't have put that video up there," he says — although he stops short of expressing remorse for spying on his wife with a drone in the first place.

All of the aerial-surveillance developers interviewed for this book were unambiguous about how WAMI might be used to similarly nefarious ends. Ross McNutt suggested that one could use his cameras to "follow politicians from their houses to the gay bars and blackmail them." Walmart, he said, could track its employees to union meetings. Suddarth said that the technology would be "a stalker's dream."

But they all mostly have little patience for the idea that the American government would actually use the technology to intrude on citizens' personal lives in those ways. McNutt, who claims to be a libertarian — a somewhat puzzling political position for a man who sells surveillance equipment to government agencies — told me, "Sometimes people think, 'Hey, we're a big deal, you're going to follow us, figure out who we are, and send the police after us.'" He went on, "I'm sorry, but I don't have time for that. If you're not killing people, not burning houses down, not shooting people, not robbing people, not kidnapping people, not hit-and-running people, I don't really care. Because that's all we have time to look at."

"The fact of the matter is," Meermans told me the first time we spoke, repeating a line that he has honed from years of operating in the surveillance industry, "the US intelligence community has the ability to spy on you, but it does not, because . . . and here's the secret, listen closely" — he paused for effect — "the US intelligence community *doesn't care about you!*"

In *Enemy of the State*, Thomas Brian Reynolds, the fictional deputy director of the NSA, echoes this sentiment. "Look," he tells a privacy-conscious senator who refuses to support a broad new surveillance bill, "I don't care who bangs who, what cabinet officers get stoned." All he cares about are the people trying to hurt Americans. (Ironically, Reynolds then proceeds to kill the senator for refusing to support the bill.)

When I asked the WAMI creators how they would feel about living under an all-seeing aerial eye themselves, their responses were uniform: no problem. "I have nothing to hide. I have no issues with anything that is going to provide some level of protection and defense of the country and where I live," said DARPA's Rick Nichols, who helped steer the ARGUS program for nearly a decade. "They're going to overlook what we're doing. Privacy concerns are there for a reason, but the government needs to have access to certain pots of information to make sure that the bad guys aren't planning ill will."

Only one of the figures offered a dissenting view: the Lawrence Livermore engineer Sheila Vaidya, who retired in 2018. Because WAMI would pick up personal information that has nothing to do with criminal activity, like the route she drives to her surfing lessons on the beach, it reminded her of the NSA's mass collection of telephone records, which came to light in 2013 via the Snowden leaks. "The fact that you are making a call to your grandmother, do they have the right to know that?" she asked, without offering an answer. "I'm not saying it's a good or a bad idea," she said. "I could live with it, but I don't want to embrace it." In an earlier discussion, she said that she would rather the technology be restricted to border security and certain limited homeland-security operations.

Another common argument is that we already lost our privacy long ago, so the aerial eye doesn't really matter. Anthony Hoogs, the Kitware engineer who is developing behavior-detection systems, pointed out that cell phone data can already be used to track us in our day-to-day routines far more closely and accurately than visual surveillance systems. Bill Hoffman, the founder of Kitware, said that he once reviewed a map of cell phone usage from an urban center and noticed large groups of devices speeding down the runway of the local airport — people ignoring the "no cell phone" rules enforced by commercial airlines.

If you walk through a public space or into a store, you're on a CCTV camera. If you've ever been arrested, even if you were never found guilty of a crime, your face may be in a federal facial-recognition database. Over the course of reporting, I was told numerous times that the real villains I should be focusing on are Microsoft, Facebook, and Google, all of which have proven to be relentless in the collection of ever more detailed information about their users. When we're being watched by so many entities already, what difference does one more eye make?

If anything, the creators say, the fact that we have become accustomed to life under watch is strong evidence that people will inevitably grow comfortable with the idea of being watched from above, too. Someday, they say, these concerns won't loom so large anymore.

Paul Boxer, the founder of Sentient Vision, the Australian automated-surveillance startup, wagers that if a wide-area surveillance system were deployed over Melbourne, where he lives, the city's population "would adapt, become apathetic, and wouldn't care."

If this was the case, I asked, why hadn't the technology been adopted already in a city like Melbourne? Because, he explained, the government isn't calling for it, and "it's probably seen as being too advanced, and too much like *1984* and the government having too much in the way of surveillance power.

"But it'll happen," he added breezily.

Laws and Order

This all being said, proponents of WAMI do generally concede that it and other surveillance technologies should be used only in accordance with certain rules. "There has to be a discussion," Bill Ross told me. "There has to be some level of regulation."

But the group is generally evasive about what these rules should be. For Ross, any regulations we do adopt need to pass a kind of Goldilocks test (my term, not his): "It can't be, 'Oh, we're never going to do [WAMI] over the US,' because that's not even realistic," he said, "but it can't be the other extreme, either. I don't think it can be the complete Wild West."

Antoniades said, vaguely, that the government should "come up with legislation about the uses of this information that we shouldn't have, just like every other evidence that is presented in a court." Meermans said that existing US laws would "probably" need to be tweaked if the police wanted to use the technology for targeted persistent surveillance of particular individuals. Crawford would only say that law enforcement agencies should have to go through a subpoena process if they wanted to fuse wide-area surveillance with other sources of data.

For some, this vagueness appears to be a matter of company policy. When I asked John Marion, who became president of Logos Technologies in 2013, what limitations he'd support, a media relations consultant sitting in on the call interrupted before Marion could offer an answer. "Because we're the contractor, we leave it for the customer, the police, the government to come up with rules," said the consultant, who asked that his name be withheld. "We can broadly say that there should be a public discussion about this, and the police are going to have to come up with some specific rules," he continued, "but we don't really make up the rules with regard to how the system is used." (Minutes earlier, Marion did say that he "strongly disagreed" with Baltimore's decision not to engage with the public before launching its program.)

Ross McNutt appears to be the only individual within the wide-area-surveillance community who has come up with specific privacy rules. His company, Persistent Surveillance Systems, maintains a strict internal policy for all employees, which is attached to any contract the company signs with a public agency.

McNutt admits that these policies are partly about protecting business. "We think we could have a pretty good value proposition, but people worry about the cost, the cost of privacy," McNutt explained when I visited the CSP operation in Baltimore, which was still secret at the time. "So by putting in the privacy policy, we lessen the perceived downside."

Nevertheless, many of these rules are reasonable. To wit: analysts must refuse all requests for surveillance that are not directly related to an investigation or a call for emergency service; in order to conduct persistent surveillance of a single location, an analyst must seek permission from a supervisor; the company does not use infrared surveillance cameras without a warrant.

As a general rule, PSS also embraces a philosophy of radical transparency. "I would like to invite all of the newspapers in here to show them what we do," he told me. "I want them to know what we do. I want them to see the imagery and say, 'I'm not worried about that.'"

12

A Murder in Baltimore, Redux

ROSS MCNUTT AND I spoke several times in the weeks after I visited Persistent Surveillance Systems' surveillance operation in Baltimore in the summer of 2016. He told me repeatedly that he disagreed with the city's decision to keep the program a secret. It would later emerge that McNutt had been emailing his primary contact at Baltimore Police, Lieutenant Sam Hood, urging the department to announce the existence of the program as soon as possible.

Each time we spoke, McNutt sounded more anxious. He decided that if the BPD didn't make its announcement by the last week of August, he would take matters into his own hands. A reporter from *Bloomberg Business Week*, Monte Reel, had heard about the program. McNutt had brought him in to visit the analysis center, and he was preparing a major feature. McNutt invited me to run a short news story revealing the existence of the program before Reel went live with a more in-depth profile.

On the morning that the news was set to break, McNutt had made a final plea to Lieutenant Hood, urging him to make an announcement before I ran my story. Hood did not respond.

McNutt told me when I was in Baltimore that he would prefer it if the public didn't think he had something to hide. But that is exactly what it looked like when *Bloomberg Business Week* published its 4,000-word profile of the operation. (Having caught wind of my story, which I had placed at *Wired*, the editors at *Bloomberg* ran their piece ahead of schedule, scooping me by a few hours.) Most city council members and state legislators only learned about the program from the *Bloomberg* story. The city's public defender had been previously unaware of the program, as had the state attorney, Maryland's congressional representatives in Washington, DC, and privacy advocates like Jay Stanley, with whom McNutt had previously been so open. Baltimore's mayor, Stephanie Rawlings-Blake, knew about the operation, but had only been briefed on its existence several months after it began.

The reaction to the news that the city had been secretly spied on for nearly nine months by an all-seeing surveillance plane was swift, negative, and ultimately fatal to the project. Trust between the public and its supposed guardians was already at a historic low. The city was still reeling from the Freddie Gray protests the previous year, and the BPD had been under investigation by the Department of Justice for discriminatory policing practices. "It wasn't the right time to do it, probably," Don Roby, a former Baltimore police officer, told me.

State and city representatives were furious. "I'm angry that I didn't know about it and we did it in secrecy," City Councilman Brandon Scott told the *Baltimore Sun*. "We have to be transparent about it and we have to make sure that we're using it in the right way, especially given all of the things that have come out about the police department."

"Widespread surveillance violates every citizen's right to privacy," Paul DeWolfe, Maryland's public defender, told the paper. "The lack of disclosure about this practice and the video that has been captured further violates the rights of our clients."

None of the individuals implicated in CSP's investigations was aware that they had been scooped up in a surveillance dragnet the likes of which no American city had ever seen. Carl Cooper, a suspect in the shooting of an elderly couple, had not been told that detectives had secured a warrant for his arrest using footage from the all-seeing camera. Neither had Kevin Kemp, who was charged with a series of crimes after McNutt's analysts tracked him riding around the city on a dirt bike for nearly two hours.

Jay Stanley pointed out that the program was violating the BPD's own CCTV data-retention policies. The ACLU of Maryland issued a statement calling on the city to end the program immediately and "prohibit use of this surveillance technology."

Congressman Elijah Cummings, whose district includes much of the city, summoned the police commissioner, Kevin Davis, to his offices. According to Cummings, Davis "apologized profusely" for keeping the program secret for so long, before launching into an explanation of its crime-fighting benefits. Cummings agreed that it could potentially reduce crimes, but he stressed that such an operation should be allowed to go forward only with the support of the legal community, civil liberties groups, the religious community, and the public as a whole.

Some Baltimore residents, who had been unwitting bystanders to the operation, were similarly shocked and outraged. One man told a video crew the day after the news broke, "We've seen what this kind of behavior leads to in the past, especially in other nations, and when it gets to this point it ceases to matter that you're actually a criminal." He added, "I hate to say that I am scared, because that's exactly what they want. Fear keeps people docile."

As McNutt had predicted, however, some members of the public welcomed the news. Another resident said that although she felt "a little violated," she generally approved of the initiative. "I think there's more crime in this city than the cops can keep up with, and I don't know if it's such a bad thing, because there are a lot of sketchy people doing a lot of sketchy things in this city." Councilman Scott noted that although he disapproved of the program's secrecy, his constituents were constantly asking him for more CCTV cameras on their blocks.

Post Mortem

By and large, those who expressed approval for the Baltimore surveillance program received relatively little airtime. In the face of the backlash, BPD suspended the daily program immediately, keeping McNutt's team on call solely for large events, such as the Baltimore Marathon.

Two months after the program was revealed, the Judiciary Committee of the Maryland General Assembly convened a hearing to discuss the city's use of advanced surveillance technologies. CSP would be the main topic of discussion. T. J. Smith, the Baltimore Police Department's spokesperson, addressed the chamber on behalf of the program.

Smith presented a passionate defense. There had been 255 homicides in Baltimore since January 1 of that year (2016), and the open-case rate was hovering at around 62 percent, he said. "We've been tasked to be creative. We've been tasked to do something more. We've been tasked to try to get these criminals off the street that are committing these acts at record pace."

The surveillance appears to have had a measurable impact. Over the course of the program, there were 21,243 calls to 911 inside the surveillance plane's coverage area. In total, McNutt's team submitted investigative briefings for 105 crimes, including 5 murders, 15 shootings, 3 stabbings, 16 hit-and-runs, and 1 sexual assault. In shootings and murder investigations alone, the company had tracked 537 targets, and had identified 73 people and vehicles thought to be "primary" suspects in these incidents.

All told, leads collected by PSS had helped investigators advance at least 10 shooting investigations. In the murder case of Robert McIntosh, the 31-year-old father of three who was shot in Madison Park, the analysts' aerial investigative work had helped lead to the arrest of a 28-year-old man named Deonta Turner, the presumed triggerman. The BPD confirmed that without the eye in the sky, the case would have almost certainly gone cold. A similar shooting had happened the same week that Turner was arrested, and because the PSS plane wasn't flying at the time, the investigation had stalled. Smith said he had been receiving calls from family members of other murder victims wanting to know if the plane had been flying when those homicides took place, too.

The surveillance program's deterrent effect appears to have been even more pronounced. In the final two months of the program, there had been an average of six daytime shootings in Baltimore every week, Smith said. In the week after it was revealed, there was just one.

Smith also raised the specter of domestic terrorism, noting a recent string of bombings in New Jersey and New York City. "If the technology is up at the time someone decides to carry out something like this, then it allows us to go back and take a retrospective look at who could have done this, and potentially taking us to a terrorist cell. And I don't think that's extreme in 2016 America to say something like that. Because we all know that they exist."

Testifying alongside Smith was David Rocah, a senior staff attorney for ACLU Maryland. In his view, Smith's insistence on the positive impact of persistent surveillance was both misleading and misguided. In an interview with the *City Paper* a few days after the hearing, Rocah said, "The concerns that we've raised about the technology is not that they are useless. It would be 'useful' if the police didn't have to get a warrant. It would be 'useful' to the police if they didn't need reasonable suspicion." But if "usefulness" was the only criterion by which to judge a surveillance technology, he remarked, "what the hell is the point of having a Fourth Amendment in the first place?"

As Rocah looked on skeptically, Smith assured the panel that the technology wasn't capable of the kind of privacy violations that would concern most citizens. He brought up a series of PowerPoint slides showing screenshots of the camera footage. "Certainly can't see Will Smith on top of a building. Can't see his face," the BPD spokesperson said, referring to the actor who played the central role in *Enemy of the State*. "As you can see," he had said, "you can make out buildings, you can make out vehicles, but you certainly can't make out people." (This wasn't strictly true. During the investigation into the murder of Robert McIntosh, as well as many others, the analysts had tracked several people on foot.)

When Representative Charles E. Sydnor asked why there weren't more crimes solved, given that the all-seeing eye appeared to be so good at catching criminals, Smith noted, again, that the technology wasn't entirely all-seeing. Lor Scoota, a well-known local hip-hop artist, had been murdered 100 yards outside the field of view of the camera, Smith said. "We only capture about 32 square miles."

Describing why the surveillance did not require a warrant, Smith told the committee about the gap in US law that leaves citizens open to aerial observation. "This is public space," Smith asserted. "You can go outside with your camera right now and take pictures as much as you want," he explained, "and it's not against the law"; the air is no different.

The following month, the Baltimore Police Department published a "Memorandum of Law in Support of Constitutionality of Wide Airborne Surveillance," which appears to be based on the legal memo that McNutt had shown me when I visited Baltimore in 2016.

Citing three Supreme Court cases that serve as the basis of US aerial privacy-protection precedents, the memo asserted that "photographs taken from a manned aircraft flying within publicly navigable airspace do not constitute a search, and do not run afoul of the Constitution." The memo also asserted, like McNutt's memo, that a multimillion-pixel military-grade camera constituted a "publicly accessible" technology, declaring that "the cameras are available to, and routinely used by members of the public."

The memo further asserted that "no infrared, telephoto, or zoom lenses are utilized," and yet in the hearing, Smith seemed confident that, once the controversy had died down, CSP would grow into a "24/7 program." "There is infra-red available, of course, that you could do something like this through the night."

"But you would want to run it through a commission first?" one of the representatives asked.

"Of course," said Smith.

In a report published in January 2017, the Police Foundation, which had agreed to serve as conduit for the donation that funded CSP on the condition that it be allowed to conduct an evaluation of the technology's effectiveness, concluded that "persistent surveillance holds potential for helping solve crime." But ultimately, it too left a number of lawmakers unconvinced.

While the authors of the report wrote that the foundation "highly recommends that a rigorous evaluation of persistent surveillance be conducted before American policing employs it on a wide scale basis," they defended the BPD's decision to move ahead with the program without seeking broader approval:

> The police do not always have the luxury of waiting until research yields scientific evidence about the efficacy of a particular approach. When people are dying the police must act to stop the violence — even when doing so carries a degree of political risk.

The authors felt that the Baltimore Police Department's decisions "to place its personal and professional self-interests aside" and push forward with a program that might not have been all that popular with the public represented "the hallmark of courageous leadership."

Representative Sydnor and two other representatives introduced a bill calling for the establishment of a task force to review the state's current and planned use of novel surveillance technologies. But after a scathing report from the House Judiciary Committee, the bill was ultimately withdrawn.

Miami's Vices and Voices

The next time I spoke with McNutt, a few weeks after the hearing, he was still fuming. The outcry over CSP, he said, was "just a couple of reporters that were hyperventilating." He explained, "It was just the media that was saying, 'Oh my God, you didn't tell us.'"

McNutt told me about an online poll from *Baltimore Business Journal* in which over 82 percent of 128 respondents registered approval of the program "as long as it's keeping people safe." A similar poll from the *Baltimore Sun* asked respondents, "Should the Baltimore Police Department have disclosed plans to conduct aerial surveillance over the city before doing it, even if it put the program at risk?" Four-fifths of the 316 respondents said no.

McNutt took comfort in those numbers, and found much reason to be optimistic. He told me that he was working to replicate Baltimore's program in another city, somewhere in the South, an effort that was looking particularly promising. His confidence, as it turned out, was ill founded.

In the late spring of 2017 the Miami-Dade county government in Florida quietly published a short document describing a $1.2 million grant application that it had filed with the Department of Justice to establish a wide-area-surveillance test program over the city with Persistent Surveillance Systems. According to the operational plan drawn up by McNutt, a plane equipped with the company's camera, capable of covering up to 30 square miles, would spend 10 hours orbiting over the city every day.

The city hoped to focus the operation on Miami's Northside neighborhood, which had experienced a "precipitous" increase in crime in recent years, according to one document outlining the plan. "In this neighborhood, families rooted in the community for decades wake up to increasing crime and violence," the grant application explained, noting that it was becoming less common for witnesses to cooperate with investigators. "There are multiple shootings every week. The fabric of the community is eroding." The Miami-Dade Police Department's hope, according to a memo from the mayor to the board of commissioners requesting retroactive approval for the initial grant application, was to use the system to "maximize convictions" and "prevent crime."

Like the operation in Baltimore, the Miami-Dade program was kept from most government officials, including the members of the county board, until after the Miami-Dade Mayor's Office had submitted the grant application. McNutt claims that he urged officials to disclose the program before submitting the proposal but was told that doing so would contravene city policy. In its memo to the board, the Mayor's Office claimed that it hadn't had time to notify the members of the board prior to the grant-application deadline.

In May 2017 the city uploaded the memo to its website without announcement — hoping, it seemed, that it would go unnoticed. Within days, it was discovered by a reporter from the *Miami New Times*. Here, too, the furor was what one might also call precipitous, and the administration's efforts to defend the program mostly didn't land. "You have no expectation of privacy when you walk outside," Carlos Giménez, the mayor of Miami-Dade County, who had authorized the grant application, told the *Miami Herald*. "I have no expectation of privacy in my backyard."

In short order, a number of local officials came out against the program. Miami mayor Tomás Regalado told a local blogger that the county government had "no jurisdiction" to conduct surveillance over the city itself. "I'm certainly not going to be in favor of having silent drones flying over Homestead spying on people," Jeff Porter, the mayor of another Miami-Dade city, told the same blogger. "You can't cast that large a net. Don't spy on all of us."

Like other regional chapters of the ACLU, the ACLU of Florida issued a statement reiterating many of the concerns that had been raised over the operations in Dayton and Baltimore. A number of local and national advocacy groups published a joint letter condemning the program. "We do have the expectation that our movements will not be tracked and recorded around the clock by our local police department," it read. "Communities of color, immigrants, religious minorities, political activists, and LGBTQ people have all been disproportionately targeted by government surveillance."

Some in the community were supportive of the proposal. "At this point, anything that's going to help get the killings down, I'm going to support," one Northside resident told the *Miami Herald*. "You can be going to the store and not make it home. You could be going to the gas station, and not make it home." But there was no concerted effort to counterbalance those protesting the program, and the day before the proposal was set to go before the county board, the application was retracted. Explaining the decision, Juan Perez, the county police director, told the *Miami Herald*, "I hear the voices."

The Kill Pattern

Despite these setbacks, McNutt continues to hold an unyielding belief that someday he will fall on the right side of history. In the spring of 2017 he doubled down on his campaign to restart the Community Support Program in Baltimore. Compared to the behind-the-scenes dealings that led to the first iteration of CSP, the new effort felt more like a political campaign. He launched a website with detailed pages for law enforcement agencies and elected officials describing WAMI's benefits. He gave a series of radio interviews, and began delivering briefings at community centers around the city.

That summer, McNutt found an unlikely ally—Archie Williams, an energetic community organizer who had spent more than a decade in prison on drug-related charges. Williams had initially been outraged by the news that BPD was secretly conducting a military-style surveillance operation over the city. Nevertheless, in May 2018 he attended one of McNutt's presentations at Simmons Memorial Baptist Church, a few blocks from where Robert McIntosh was gunned down almost exactly a year earlier. McNutt recalls that Williams had spent most of the briefing with his arms crossed defiantly until, after walking through his more rehearsed arguments, McNutt had started talking about how the cameras could watch the city's notoriously corrupt police force to verify their claims regarding shootings and warrant applications. This, Williams found compelling.

During the pilot program in 2016, the camera had recorded two police shootings, and it had occurred to McNutt that the data could be used to validate the official police department accounts of the incidents. He told me that Persistent Surveillance Systems had already helped a public defender contest charges against a client who was arrested after police obtained a warrant claiming they had seen drug-related activity at his home. The camera footage showed that just two people had visited the address in the period cited in the warrant.

Engineers and officials who know WAMI often talk about being "converted" or "drinking the Kool-Aid." The presentation at Simmons marked the beginning of Williams's transfiguration. After the event, the church's pastor, Williams, and several other audience members decided to launch a nonprofit organization, Community With Solutions, that would support McNutt's campaign through grassroots activism.

McNutt and members of Community With Solutions approached a number of city leaders, including Baltimore's newly elected mayor, Catherine Pugh. McNutt was confident that the John and Laura Arnold Foundation, which had backed the original surveillance program over Baltimore, would provide enough additional funding to keep PSS airborne for a full year. According to the *Baltimore Sun*, Pugh agreed to consider approving a rebooted CSP as soon as McNutt and his Community With Solutions advocates could demonstrate that the idea was widely supported among city administrators, community leaders, and law enforcement officials. "When things bubble up from the community," Pugh told the paper the following year, "I think you have to listen."

Community With Solutions set up dozens of meetings for McNutt and Williams, including discussions with several chambers of commerce and the Baptist Ministers Conference. During a phone interview in May 2018, Williams told me that he had personally held more than 100 meetings with community organizations and city leadership. According to the *Baltimore Sun*, Williams would tweak his pitch according to whom he was addressing. If he was speaking to a business group, he might talk about reducing crime; in discussions with an advocacy organization, he would focus on how CSP would monitor police.

Both McNutt and Williams said that Community With Solutions

did not receive money from McNutt's Persistent Surveillance Systems. The company did, however, provide a web developer to build the activist group's website and produced videos for Williams and his peers to use in briefings. McNutt also allowed Community With Solutions to conduct briefings in the very same office it had used as its base of operations in 2016, and provided, according to Williams, "input in how we can come about in an impactful way." (In my conversation with Williams, I noticed that many of his arguments echoed McNutt's main talking points, often almost verbatim.) Williams also acknowledged that McNutt had supported him "personally" with items for his one-year-old son, including a car seat.

When I asked Williams about the privacy concerns, he admitted that the technology raised "policy issues," but said that McNutt would address those issues with the company rulebook. "As for me? I'm cool with it," he said. "Right now it's a kill or be killed mind-set," he told me. "If somebody comes up and shoots me today in broad daylight, they don't believe they will be caught. The changed mind-set is, 'Oh, wait a minute, I heard that plane might be running, I don't think we should do this today.'"

"That saves two people," he continued. "The person who's going to get killed and the person doing the killing."

Later in the call, Williams asked if he was still on the record. I said that he was but that we could go on background if he wanted. "No, keep me on record, because I think this is important," he said. "Anybody who has an idea to save life, whether they be pink, purple, brown, or black, should have a chance to do it."

He grew animated. "People always say, 'Well, OK, what if the cameras get better? Ross will tell you, 'Listen, if my cameras get better, I will look at more. I want to keep on being able to solve unsolvable crimes." In the background, his son started to cry. "We don't want people thinking that they can kill someone and get away with it," he said. "As long as you can disrupt that, I'm cool. I'll feel like I have accomplished something, a disruption in the kill pattern."

When McNutt and I spoke a few days later, it had been close to two years since the first iteration of the Community Support Program had come to an unceremonious end, and there was still no word from Mayor Pugh's office as to whether she intended to give it a new green light.

I asked why McNutt remained so motivated in spite of the setbacks in Baltimore and Miami. He said that in 1993, as part of a volunteer program called Fifty for Colorado, he had spent time in a rundown neighborhood in Denver: "I got to see what it was like and it pissed me off." He described witnessing a racially charged encounter between a police officer and a resident that had stuck with him ever since. "When I had the chance to see what impact the system that I built had, or could have, on the inner cities," he said, "I thought that this would make that life better."

It was clear that McNutt wasn't banking on Baltimore to make him rich. Nor was his financial security entirely dependent on the city's decision. He told me that he subsidized PSS with revenue from a successful aircraft-leasing business he owned. Not long after we spoke, he was awarded a large grant from the Ohio Research Network for a project to develop autonomous passenger drones, which would operate like aerial taxis to shuttle people from place to place.

Once the program in Baltimore was relaunched, he said, he would only charge the city at cost. "From a business standpoint, I ought to close Persistent Surveillance Systems and forget about it," he said. "But I think it has too much potential impact to help cities, and I'm not willing to give up on it."

13

Where Power Meets Fury

ON MY FIRST day in New Mexico with Steve Suddarth in June 2017, we had secretly surveilled the city of Albuquerque. We had watched students exercising on the sports field at the University of New Mexico and tracked cars darting along the downtown thoroughfares. Orbiting over Sandia Heights, I had learned the first lesson in wide-area surveillance. On the second day of my visit, I would stay on the ground and become the target.

Suddarth lives in a residential community with its own airstrip, about 30 minutes outside Albuquerque. He can taxi his white Cessna right up to the driveway of his home, which is adjoined by a hangar that is larger than the house itself. The hangar serves as the headquarters and workshop for his company, Transparent Sky, which builds a range of WAMI cameras and software systems.

The sky on the morning of that second day was perfectly clear. Suddarth would be flying in the Cessna with his assistant, a taciturn man who used to work as an electrical engineer on fast-attack nuclear submarines. I would be accompanied on the ground by Suddarth's business partner, Greg Walker; Suddarth's wife, Deborah; the company's intern, Vanessa; and the Suddarths' dog, Duke.

When the airplane reached cruising altitude, I stepped out onto the driveway with Vanessa and Duke. To the naked eye, the Cessna was nothing more than a tiny white speck against the vast desert sky. But the camera on the plane could make us out clearly, our shadows cutting diagonally across the tarmac in the midmorning sun. Over the radio, Suddarth suggested that we drive to the local Walmart a few miles away and do some tracking exercises in the store's large parking lot. I bet him a beer that he couldn't keep us in his sights the whole way there.

As Walker drove Vanessa and me in his truck (Deborah and Duke stayed home), I stuck my head out the window to try to spot the airplane, but I couldn't see it anymore. From time to time, Suddarth would call out our exact location over the radio.

The area was sparsely populated and the houses were slung low to the ground. It felt like there was nowhere we could hide. I asked Walker if we could turn down a side street and do a U-turn to try to throw Suddarth off the scent, but he demurred. He said that he didn't know the area very well and didn't want to get lost. Vanessa suggested we take the second entrance into the Walmart. Steve, she said, wouldn't be expecting that.

When we pulled up to an empty corner of the lot, I stepped out of the truck and tried, again, to spot the airplane. Again, I couldn't find it. Again, Suddarth came in over the radio; he still had eyes on me.

That afternoon, Suddarth passed the footage to a group of researchers at the University of Missouri who ran it through a com-

Left: The author standing in a parking lot at the Walmart Supercenter in Edgewood, New Mexico, as Steve Suddarth's wide-area surveillance plane (not pictured) circles 10,000 feet overhead.

Right: The author, seen through Steve Suddarth's wide-area surveillance camera, standing in the Walmart parking lot. *Courtesy of Transparent Sky, LLC*

puter-vision processor that tracked our forms in the footage wherever we went. I could have driven a hundred miles in any direction and it wouldn't have made a difference. In other words, I had finally come face-to-face with "the instrument of permanent, exhaustive, omnipresent surveillance," as Foucault had put it in his famous text on the Panopticon, "capable of making all visible."

This is where the theory of wide-area motion imagery and automation meets the reality. Where the race-against-the-clock heroics of the national laboratories, the armed services, and the intelligence agencies ends and the long road of the technology's routine use in everyday life begins. Where the engineers' resolve to make a positive impact comes up against the fact that no tool of raw power is intrinsically good or evil in its essence, but that, rather, everything depends upon how we choose to use it.

Wide-area motion imagery opens our lives to a view from the heavens that had, from the dawn of civilization until not so long ago, been reserved for the gods and the stars. It is impossible to deny that such power could be used as a force for good. If the NYPD had possessed the technology at the time I witnessed the shooting of Taekwon Hart, I would have wanted it to be used in the effort to apprehend Hart's assailant. Under an all-seeing eye, organized crime networks can be unraveled, kidnappings can be resolved, hurricane survivors stranded on their rooftops in the dead of night can be found and delivered to safety.

But there is no single widely used surveillance technology in history that has not, at some point, crept too far beyond its original purpose or given rise to unintended consequences, causing harm as a result. While we have only just begun to discover what they may be capable of, both for good and for evil, it is already clear that the twin furies of WAMI and automation have the potential, if mismanaged, to make surveillance much more invasive, unscrupulous, mysterious, and unfair. Any harm that was to arise from the technology's improper use would, in turn, undermine efforts to use it, in the words of the engineers, "the right way."

So we're going to have to establish ground rules to make sure that its potential dangers can be averted. And that begins with an honest appraisal of why the all-seeing eye, in certain circumstances, poses serious risks of harm.

The Road to Hell

The engineers who developed WAMI are right when they point out that widely embraced technologies like CCTV and social media have already laid our lives bare. But private spaces do still exist. And by design, wide-area surveillance is a tool for intruding upon those spaces. One such "space" is the track of our movements across large areas. WAMI uncovers that track and reveals both where we go and who we associate with — something that, as you now know, is difficult to achieve with any other technology that could be used without a warrant. If anything, the total encroachment upon our digital lives, in particular, makes this little privacy we have left in the physical world all the more precious.

The simple and unfortunate fact of the matter is that the history of surveillance technology is full of stories of intrusion into sacrosanct private spaces for ignoble ends. Shortly after the invention of wiretapping at the turn of the 20th century, police in New York began listening in on random calls from restaurants and poolrooms in the hopes of catching criminals who would have otherwise escaped detection; more often than not, they heard instead discussions that were none of their business. It was also not all that uncommon for police to intercept residential wires just for the fun of it. More recent tools like social media monitoring software, cell phone interceptors, visual and electronic airborne surveillance, and facial-recognition systems have all been used against US advocacy groups that were doing little more than exercising their constitutional rights.

Wide-area motion imagery represents a maximalist approach to intelligence and surveillance. It is capable of embodying, without caveat, the spymaster's mantra — Collect it all — and it, too, could creep beyond its original intended purpose and be misused if those using it are left to their own devices. From chasing sexual predators and violent offenders, it is a small jump to tracking traffic violations and illegal dumping activities, and an equally short jump to following demonstrators, members of opposing political parties, or deviant religious groups.

In part this is because, in the absence of clear laws restricting the technology's use, what counts as fair surveillance is largely a matter of opinion. A few hours after Suddarth tracked me to the local Walmart in New Mexico, he and Deborah took me to a grungy roadside bar at the base of Sandia Peak. It was still gorgeously sunny, and we sat outside on the bar's terrace overlooking a small gravel parking lot filled with Harleys and large pickup trucks. After buying him the beer I owed him from our wager, I asked Suddarth to tell me why he thought we shouldn't be afraid of the prospect of domestic wide-area surveillance, or, as he at one point somewhat jokingly called it, "toutveillance," the surveillance of everything.

"If you're insecure," said Suddarth, who was wearing a DON'T TREAD ON ME hat, "there's nothing else that matters in your life but making sure that you are secure." If, on the other hand, you had security, he explained, you could engage in higher, more productive pursuits. Liberal democracy, he said, savoring his beer, was predicated on security. "People can only talk about privacy because they feel secure. If they feel insecure, they're going to drop the privacy argument pretty fast."

To illustrate his point, Suddarth offered what he warned me probably wouldn't be taken as the most politically correct example. India, he argued, is only a successful society today because of several hundred years of colonial rule that had imposed a strict order upon the country's complex patchwork of historically adversarial ethnic groups. Suddarth believed the ultimate purpose of Angel Fire was, similarly, to support what he described as the US military's "real reason" for being in Iraq; he took General Abizaid's 2003 memo warning the Pentagon about the threat of IEDs to mean that quashing the IED problem was, like British rule in India, key to establishing an order that would put a lid on the millennial ethnic and tribal alliances that formed the basis for the country's unrest. Imagined or real, it was an objective that he appeared to have been happy to get behind.

The previous day, Suddarth had told me that in the summer of 2015 he had flown his surveillance plane to Ferguson, Missouri, to orbit over the protests and riots that erupted after the fatal shooting of an African American teenager, Michael Brown, by a white police officer. The goal of the operation, he had told me, was to collect sample footage to test automated tracking and analysis algorithms. Even so, the technology gave Suddarth astonishing power over those on the ground. He said that if he had recorded any suspicious activity, he would have passed the information on to local law enforcement authorities. In other conversations, Suddarth, who is white, and Deborah, who is black, had told me that they fervently oppose the Black Lives Matter movement. Steve referred to some of the protesters in Ferguson as "thugs."

Suddarth was not alone in appearing to allow his politics to seep into his work. Ross McNutt, who had told me that political surveillance was one of the foremost perils of WAMI, said that Persistent Surveillance Systems "doesn't do protests," but then conceded that the company would fly over "a World Bank type thing" if contracted to do so. He explained that his analysts would focus only on "agitators" — the "20 idiots who are out there to cause anarchy."

Suddarth and McNutt never gave me any reason to believe that they don't genuinely want to make the world a more peaceful and harmonious place for everybody. But of course, to the British Raj, Mahatma Gandhi was an agitator trying to cause anarchy, and the colonial occupiers would have undoubtedly used an all-seeing eye to identify the participants in Gandhi's Salt March, just as the FBI would have used a Vigilant Stare over the Selma-to-Montgomery March in 1965. Those conducting the surveillance would have been certain, like Suddarth and McNutt, that they were serving the just cause of peace and security, but where would we be today if the watchers had had their way?

The Dragnet

Of course, surveilling peaceful demonstrators and political enemies is an egregious abuse, and any true democratic society is already enshrined with rules to protect these groups from intrusion, even if those rules aren't always followed. But even the constrained use of wide-area surveillance could have subtler, if no less troubling, corollaries.

Because wide-area surveillance collects it all, and because everybody can be perceived as potentially suspicious when they are little more than a single pixel from 10,000 feet in the sky, the all-seeing eye will push up the number of encounters between law enforcement and citizens. While in many cases this means more criminals will be found, WAMI will also sometimes lead police to innocent citizens who merely happen to make a U-turn in the wrong place.

That could be harmful to the relationship between law enforcement and the general population, a relationship that is already fraught. Worse still, the technology could engender a style of relentless activity-based intelligence that treats all individuals as unknown unknowns — possible criminals who can only be discerned through persistent surveillance.

This would be hard to reconcile with the community-based policing that some regard as necessary to salve law enforcement's relationship with communities in the areas where it has been most damaged — areas like Baltimore and Ferguson, where WAMI is also most likely to be used for crime fighting.

At one point during the 2016 CSP pilot program, Baltimore began using the technology to identify people who had witnessed crimes but hadn't come forward to provide testimony, in some cases tracking eyewitnesses all the way back to their homes. Though the tactic probably wasn't technically illegal, it was hard to see how such a coercive strategy wouldn't further alienate the city's residents or chip away at the principles that those same officers are supposed to protect. The right to refuse to provide testimony is exactly that: a right, and an inviolable one, to boot.

Even if a law enforcement agency used WAMI strictly in keeping with the kind of narrow and specific rules laid out in the following chapter, that does not mean that the power to surveil would never be made available to someone holding no regard for those rules. Every analyst with the system password would have access to detailed personal information revealing thousands of citizens' routines, social networks, and home addresses. Some of those analysts might, perhaps, struggle to resist the temptation to spend a few minutes — or hours, or days — tracking a spouse, say, or a celebrity, or a local politician.

And because WAMI is particularly useful in forensic investigations, departments will want to hold on to the data for a long time in repositories that may or may not be all that secure. Here, the people with the password are less of a problem than the people who know how to bypass a password by illicit means. Local authorities have never been known for their strength in cybersecurity, and if a member of a white supremacist group hacked into a dataset like Suddarth's footage of the Michael Brown protests, they could potentially track demonstrators to their homes. Unscrupulous political consultants might use WAMI footage for opposition research. Foreign regimes might track all the employees leaving a sensitive research facility, the way the Livermore engineers had hoped to use their planned surveillance satellite to follow North Korean nuclear physicists commuting from work.

Given how, as we saw in chapter 10, much of the information wide-area surveillance generates is only actionable when cross-referenced against other sources, its use will also drive up the collection of other forms of surveillance that are themselves exploitable and subject to abuse. In Baltimore, for instance, the use of WAMI in the McIntosh murder investigation prompted the police to co-opt the camera on a city bus and use it as a license-plate reader.

The theory of activity-based intelligence posits that it is always advisable to collect as much data as possible and hold on to it as long as possible in case it becomes relevant in the context of new information from other sources; by way of example, Dave Gauthier, the National Geospatial-Intelligence Agency's senior official for ABI, suggests that one might wish to record every license plate in Baghdad, seeing as how "some of them will be relevant in the future when an event happens." This so-called "mosaic theory of intelligence" has served as the basis for the dragnet techniques that largely define today's surveillance paradigms abroad and — increasingly — at home.

The more effective fusion gets at linking disparate pieces of information, the more it will drive the thirst for information from alternate tools that further implicate citizens in the tangle of data-collection devices that has already intruded upon so much of our lives. This will, in turn, lead agencies to more closely scrutinize all of the data that they collect, as every piece of information, no matter how mundane, could become significant in light of another datum from a separate source. If carried to its extreme, this dogma would have just one logical conclusion: a world where everything is regarded with suspicion.

All of which brings us to the final danger of wide-area surveillance: the technology has the potential to create fear.

Make no mistake, this is a function of design just as much as it is a consequence of use. According to a 2005 Pentagon operating paper, a primary purpose of any persistent-surveillance system is to give "the impression that we can 'observe' even an adversary's very intent," so that he "is constantly looking over his shoulder, sure he is being watched, followed, tracked, and heard." Even many of the programs' names are unsubtly coded threats: the vigilant Constant Hawk, the hundred-eyed giant ARGUS, the petrifying Gorgon Stare.

Unfortunately, fear can be an imprecise weapon. Being persistently surveilled from the sky is often just as discomfiting for people who ought not have anything to fear as it is for those who do. In war zones, civilians living under drones have described inhabiting a state of perpetual unease, fearing they might become an inadvertent target of an operation. While it could be a good thing if violent criminals thought twice before stepping out into the open air, that's no way for everyone else to live.

For my part, I want to be able to make a legal U-turn and not have to worry that a surveillance system, flagging my action as an anomaly, is now tracking me as a suspect. I want to be able to buy a drink at a bodega without fearing that I may be implicated in a persistent counternarcotics operation. I want to be able to go to a political rally and not wonder whether some unsavory actor, somewhere, might be watching me from above.

It is true that with time and some attitude adjustment on our part, the fear of an all-seeing eye might subside. We might learn to steer clear of protests or political gatherings or bodegas. We'd think twice before parking our cars in majority-Muslim neighborhoods or interacting with people who, for whatever reason, have come under government or police scrutiny. We would eventually come to accept that those sacrosanct spaces are no longer ours — but that's probably the scariest scenario of all.

The Bullet and the Brain

Fortunately, none of these dangers are particularly novel. As we'll see in the next chapter, many can be protected against with measures that have already been proven to work for other technologies. The prospect of *automated* wide-area surveillance, on the other hand, raises concerns that are wholly vexing and unfamiliar.

Like regular wide-area surveillance, automated surveillance could, and should, be used for beneficial purposes. The Vision Lab at the University of Dayton is using many of the same techniques that could detect undocumented migrants crossing the border to scan airborne imagery for construction machinery that might be digging too close to submerged pipelines. In a test in a South African wildlife park in 2018, a drone equipped with an AI system was capable of finding poachers far more quickly than regular human observers. Kitware, the computer-vision company, has retooled some of its WAMI software for the Human Connectome Project, which is working to build a complete map of human brain connectivity.

But again, those applications will be undermined if automation's darker possibilities go unaddressed and harm arises as a result. This is particularly true for law enforcement, which, like the Pentagon, has already embraced a range of automated technologies. As of this writing, dozens of cities in the United States use predictive-policing software that runs historical crime data through sophisticated algorithms in order to identify the likely locations of future crimes. A number of states and jurisdictions also use algorithms to automatically generate sentencing decisions in certain criminal cases. Many law enforcement agencies are actively seeking to incorporate facial-recognition and fusion capabilities into CCTV surveillance systems and even bodycams. Much of what you're about to read applies to all such forms of automation in policing.

A foremost concern among privacy advocates with regards to automation is that in a city where an eye in the sky — or every surveillance camera on the ground — is linked to an automated behavior-detection algorithm, we all become *actively* watched.

Good surveillance algorithms like the ones being developed by the Pentagon are, by nature, alert and suspicious creatures. They regard every piece of information presented to them as a potential *positive* that must either be confirmed or dismissed. Good cops are this way, too. And to be sure, many of the same behaviors that lead an algorithm to target an innocent citizen would likely spark a human officer's suspicion as well. But unlike CCTV cameras, there aren't cops on every street corner. For every time a cop pulls someone over because they have misjudged an innocent action, there are thousands of people engaging in that exact same behavior with no cop to see them.

An artificially intelligent all-seeing eye would see *every* anomaly. Every behavior within the camera's view would be regarded as a potential threat by a mysterious and inflexible algorithmic logic that we don't truly understand. Many automated surveillance systems for WAMI already come with a "trip wire" feature, which automatically tracks every vehicle that passes through a designated area of interest. In the battlefield, this is useful for keeping tabs on activities around a remote network node. In a domestic setting, this means that the eye in the sky will track you for simply driving along the wrong block.

It doesn't take much imagination to foresee how being actively watched at all times by powerful robotic surveillance apparatuses might, just might, further erode the trust between the surveilled population and those charged with protecting them — but the dynamic between the people and their overwatchers is not the only delicate relationship we have to consider. We also need to think about what goes on between the automated surveillance system and the human analyst herself.

Imagine that an Air Force targeting specialist is monitoring a drone operation when an automated activity-detection program informs her that an individual who has been known to carry out attacks on US targets is driving a vehicle packed with explosives toward a contingent of US soldiers on the ground.

The targeting specialist wants to be sure that the computer is positively certain that it has, for lack of a better term, "positively" identified the target. If the computer is wrong—and computers are often wrong—the targeting specialist, who might have only moments to make a decision, could be ordering a strike on a civilian. So, if the computer is only *sort of* sure, it needs to be able to convey that information to the human. But how do you quantify the degree to which a computer is certain that it is right?

Many automated detection systems that interact with humans employ a scoring system, usually presented in the form of a percentage, that quantifies the extent to which the computer trusts its own conclusions. When Livermore's Persistics software tracked suspicious vehicles in Gorgon Stare footage, it would give the probability that the car at the beginning of the track was the same vehicle in the crosshairs by the end of the mission.

In one screenshot of a Project Maven algorithm operating on imagery from a harbor, a boat is indicated as a "boat" with a score of 85 percent. A large warehouse is labeled as a "building" with 90 percent confidence; a strange hangar-like structure is also marked as a "building," but only at a confidence rate of 40 percent. These scores might be based on factors such as the quality of the footage and the degree to which the full sequence of the activity of interest is visible in the frame.

Google's online computer-vision service, Vision API, employs a similar technique, and it's worth trying it out for yourself. When I visited the site in 2016 and uploaded an image of a futuristic military drone prototype called the X-47B, the software concluded that the picture contained an "aircraft" with a 92 percent confidence level. It was less sure (85 percent) that the aircraft in question was specifically an "airplane," and only 71 percent certain that the X-47B was a military aircraft. It also concluded that it was "very unlikely" that the image contained anything pornographic or violent.

The computer took longer to process Jackson Pollock's *Untitled. C. 1950*. It was 85 percent confident that it was looking at art. (Its prejudices seem to be skewed toward the classical canon: when I gave it Botticelli's *The Birth of Venus*, it was 93 percent sure the image was gallery-worthy.) It identified the presence of "black and white" with an 83 percent confidence level (which is oddly low, when you look at the actual painting). It would only say that it was "unlikely," rather than "very unlikely," that the picture contained pornography or violence.

Such scoring systems are already common in domestic law enforcement. For instance, the automation company SeeQuestor's CCTV-analysis program rates its results on a scale from "likely" to "unlikely."

These scores, and the disparities between them, can be a matter of life and death. Say you are a police officer supervising a computer-vision system attached to a wide-area camera feed of a city. Twenty minutes after a violent armed robbery, your automated tracking system highlights a vehicle that it believes fled the scene of the crime. It is 71 percent certain of the result.* Do you send patrol cars to intercept the vehicle and tell the officers to expect an armed encounter? Or do you second-guess the computer and verify the track yourself, potentially delaying the capture of the criminals?

The question is even more fraught when it comes to predictive detection software. In a paper titled (rather unoriginally, I might add) "Eye in the Sky," researchers from the University of Cambridge, India's National Institute of Technology, and the Indian Institute of Science described a computer-vision system they had built that could detect violence in footage from consumer drones with high accuracy. What if such a system concluded, with a 71 percent confidence rate, that an armed assault was *about* to take place on the corner of Lewis and Van Buren? If the computer is right, you could prevent a terrible crime.

But if the computer is really just detecting a group of children playing on the street,† you will be putting those kids in mortal danger by introducing them to a number of armed police officers who may be expecting a violent encounter.

---

* Some facial-recognition products on the market present operators with a list of several likely suspects — like a digital police lineup — in place of a single score, but that does not necessarily solve the problem; if anything, it could make it worse, since it implicates individuals who ought to be kept entirely above suspicion.
† In a clip of test footage from the "Eye in the Sky" study, the actors who are pretending to fight are almost laughably unconvincing. Some even look like they're having fun.

On the battlefield, misplaced trust in automated systems has already led to tragedy. When a US guided missile cruiser shot down an Iranian passenger jet over the Strait of Hormuz in 1988, killing all 290 people on board, it was because the ship's automated radar-defense system had misidentified the airplane, which was traveling in commercial airspace, as a military aircraft, and the ship's crew had not sufficiently questioned the computer's analysis before approving the missile launch. In 2003 a similar error led US forces in Iraq to shoot down a British jet fighter.

Insufficient trust is equally problematic: more aviation accidents than you might think are caused by human pilots ignoring automated alert systems in the cockpit. Say, in another instance, that the law enforcement computer is only 60 percent sure that it has detected an imminent assault. That seems low, but would you ignore it, given the consequences of doing so if the computer turns out to be right?

Because breakdowns in trust can result in what Dr. Greg Zacharias, the Air Force's chief scientist, described in a lecture in 2016 as "high-regret actions," the relationship between a human and a computer can be very fragile. If a computer makes an *inaccurate* detection with a 95 percent confidence rate, the next time it offers a high-confidence analysis it is unlikely to be fully trusted by its human supervisors. Its algorithms can be tweaked, but that wouldn't prevent different types of mistakes that are equally difficult to anticipate. Once it is lost, this trust is hard to regain. After Constant Hawk imagery analysts noticed that the program's automated tracking features sometimes made errors, they stopped using the software entirely and reverted to brute-force tracking.

On the other hand, if humans ignore a computer's 51 percent analysis, and it turns out the computer had been right, the next time that it reaches a 51 percent conclusion, they are likely to overly trust it.

It will be a long time before any kind of consensus is reached as to how to account for these issues in either the military domain or the civilian world. Many countries' rules of military engagement require that forces be reasonably certain that the target of a planned attack is a military entity. But no international body has adjudicated on whether a score of 71 percent counts as reasonable certainty, reasonable *un*certainty, or just a random number on a computer.

As of 2018, after half a decade of formal debate about *lethal* autonomous weapons — machines that can find, identify, and kill human beings without any human operators in the loop — the international community hadn't even agreed on how to define what we mean when we talk about killer robots, let alone automated video-surveillance systems, which had already begun proliferating widely.

Present-day civilian legal systems are just as unprepared to deal with these trust issues. If a 71 percent confidence decision from a computer leads to an encounter in which an innocent man is killed, who is responsible? Is it the police officer in the command center who trusted the computer and dispatched a team to the address? Is it the police officer on the ground who pulled the trigger? Is it the manufacturer of the system? Or is it the police department?

What if a family member of a murder victim sues a police department for ignoring a 71 percent confidence prediction of the assault? Would it be equivalent to a police officer ignoring a warning from a human colleague? Would the legal balance shift if the confidence score was higher or lower? Investigators might run the data through a different automated system and find that it produces a completely different score — what then?

Not a Game

The issue at hand is not just that computers are sometimes wrong. It's that they can be wrong in strange and mysterious ways. When IBM's Watson, an artificial intelligence program, appeared on the quiz show *Jeopardy!* in 2011, it crushed two of the most successful human players in the game's history, but it also made some mistakes that neither of its competitors would have made even if they were on acid.

When presented with the prompt, "In May 2010 5 paintings worth $125 million by Braque, Matisse & 3 others left Paris' museum of this art period," Watson responded, with 97 percent confidence, "Picasso."

It doesn't take a degree in art history to know that "Picasso" is not an art period. Watson had become confused because the question was asking for a particular Paris museum, and Paris does, indeed, have a Picasso museum (the correct answer was "modern art" — neither of the two human contestants got it right, but they were also far less confident of their answers than Watson).

Although it is usually fairly easy to predict human mistakes (one of Watson's opponents answered, very reasonably, "impressionism"; the other "cubism"), it is still all but impossible to predict exactly what type of mistakes a computer might make. When I showed Google's vision system a bowl of apple slices, it informed me that the image was of a "flowering plant."

Police officers sometimes act in crazy and unpredictable ways, too, but humans generally appreciate the stakes of their decisions. Most police officers understand that mistaking a banana in a driver's hand for a cell phone is not the same as mistaking it for a gun. In one case, the stakes are low; in the other, they are high. Even if an officer might sometimes fall victim to passions that cloud his or her judgment, that same capacity for passion is what gives the officer the solemn sense of responsibility that acts as a check on his or her decisions, especially when those decisions fall outside the scope of what's ordinary.

Software is more like a police dog that knows how to detect drugs but doesn't understand that the drugs are illegal. Just as a police dog doesn't know that its actions might send someone to jail, a computer could not comprehend that its confidence level might be a matter of life or death.

Nor do the disinterested analytical-logic sequences of a computer's algorithms necessarily countervail the human biases of police officers, both conscious and unconscious, that, for instance, make black and Latino drivers far more likely to be searched and arrested during traffic stops than white motorists. A growing body of research, as well as a few frightening anecdotes, suggests that algorithms do precisely the opposite.

Take the Nikon Coolpix S630 digital camera, which was released in 2009. One reviewer praised it for its "excellent design" and "very good feature set." Among those features was a simple computer-vision system that alerted the user if it detected that someone in the picture had their eyes closed.

For most users, the system worked well. Not so for Joz Wong, a Taiwanese American. No matter how many times Wong took a selfie, the S630 would offer the same cheerfully racist message: "Did someone blink?" The camera, it turned out, had not been programmed to recognize Asian eyes.

Another, even more egregious, example: a "smart" HP webcam released the following year that appeared to refuse to recognize certain black users' faces but had no trouble locking on to their white counterparts. More recently, in 2017, the CEO of a company called FaceApp had to apologize after the app's "hot" filter, which employs neural networks to make your selfie more attractive, appeared to consistently lighten the skin tone of darker subjects.

Presumably, the engineers who built these systems did not intend to offend people of color with their work. But because computers behave less like foils to our passions and more like extensions of our socialized—and biased—brains, they can perpetuate, rather than alleviate, bias and discrimination. And because they are unfettered by the emotional and intellectual counterbalances (not to mention the desire to keep one's job) that might hold a human brain's prejudices in check, they will do so with unrelenting consistency and abandon.

An FBI-sanctioned study from 2012 found that facial-recognition systems were "consistently" less accurate for women, blacks, and young people, a fact made all the more troubling when one considers that 80 percent of the tens of millions of Americans in the bureau's mugshot database have never been found guilty of an offense. An in-depth study by ProPublica found numerous cases of automated sentencing programs handing down harsher punishments to black citizens than to white people who committed the same crimes. Biased outputs have been detected in algorithms used for everything from calculating credit scores to estimating a person's employment eligibility.

It's easy to foretell how similar issues might crop up in a system designed to autonomously detect behaviors associated with serious crimes. Something as simple as the extent to which a particular neighborhood has been surveilled can lead algorithms to treat an area with elevated suspicion, even when the actual recorded crime levels in that area are low. In one simulation that explored automated bias in policing, a group of RAND researchers found that a crime-prediction algorithm trained on historical data tended to characterize neighborhoods that had been subject to more surveillance in the past — a factor that is often a function of bias in real life — as being more "risky" than less surveilled neighborhoods, despite the fact that the crime rates across those neighborhoods were identical.

If a system builds normalcy models fueled by historical data, it may be more likely to characterize behaviors as suspicious in poor neighborhoods that have traditionally been more surveilled, especially if the models on which it is based are sourced from data taken from more privileged "normal" neighborhoods. A wide-area surveillance computer might simply be more inclined to flag a U-turn on the "wrong" side of the tracks as suspicious than a U-turn on the "right" side.

If a biased wide-area system were fused with other sources of data generated by other automated programs that themselves might be biased, the problem could be made worse. Say, for instance, that a WAMI feed is fused with information about the criminal records of each neighborhood's residents. Now imagine that some of those criminal records were set by automated sentencing software that tends to hand down harsher sentences to offenders of a certain ethnic minority.

If the WAMI system holds a lower bar for suspicion for those neighborhoods with harsher sentencing averages, it will lead police to more flagged anomalies in that neighborhood than in the adjacent neighborhood, where there are fewer sentences. This will lead to more arrests in the more surveilled neighborhood, which will further fuel the bias in the automated sentencing program. What is particularly frightening is that such algorithmic echo chambers could spring up and perpetuate bias without anybody noticing it until real damage has already been done. There's a reason that some in the field have taken to calling certain algorithms "weapons of math destruction."[*]

But we are not on the brink of a surveillance-based doomsday, yet. We have a little time to act — and act we will.

---

[*] A phrase coined by the statistician Cathy O'Neil in a must-read book of the same name.

14

Rules for the Eye
and the AI

FOR A COUPLE of decades following the introduction of wiretapping in the early 20th century, police generally didn't need a judge's permission to listen to private phone conversations. This is not the case anymore, because the US legal code prohibits warrantless tapping. If a police department were to use an infrared camera to record me in my home, I would have recourse to challenge its actions in the courts, thanks to a Supreme Court ruling that restricts that tactic. Likewise, there's a reason the CIA appears to have tested its surveillance systems on doctors' offices in Lubbock, Texas, rather than on a real criminal network in the city: under federal law, intelligence agencies are prohibited from spying on the domestic population for active operations.

The legislative branch also serves as a check on surveillance, if there is sufficient political will. When the Department of Homeland Security, in 2007, established a program that sought to use the intelligence community's satellites to conduct surveillance over US soil, the effort was quashed by Congress.

These checks are a natural, and necessary, response to any surveillance activity. That's because in the dogged pursuit of justice, a truly resourceful law enforcement or intelligence agent will act equally in accordance with what the law *doesn't* expressly forbid as what it does.

As of this writing, there is nothing in the US Code or Constitution, nor in the laws of most developed countries, that explicitly restricts the use of wide-area surveillance, be it in the air or on the ground, manual or automated, fused or unfused, to pursue any individual in public space. So naturally, law enforcement is going to use it. Baltimore's memo outlining the legal basis for its surveillance program, for example, focused primarily on what standing US law does not prohibit, rather than what it actively allows.

It might, in other words, be prudent to take our own cue from the film that inspired this whole story in the first place. "We knew that we had to monitor our enemies," *Enemy of the State*'s fictional congressman Sam Albert declares after the NSA's abuse of its Big Daddy satellite and other weapons is exposed. "We've also come to realize that we need to monitor the people who are monitoring them."

The window of opportunity to do so, to put the rules ahead of the technology, is getting narrower, so it would be best to act soon. Catching up with a new technology and plugging the negative regulatory spaces that it creates can take time. Police departments in the United States began publicly acquiring drones in 2009. As of this writing, nine years later, there are over 600 police departments in the country with unmanned aircraft, and yet less than a dozen states have passed legislation governing the use of the technology (ironically, those drone laws that do exist actually create a loophole for the use of WAMI, since it is generally not used aboard drones), and the federal government has done little more than issue a set of vague voluntary best practices.*

There is no reason to wait for every large city in the country to buy a wide-area surveillance system before we set reasonable limits on its use. We already know exactly how the technology works and how it would be used. We know its benefits and we can anticipate its risks. It will be much easier to set rules in advance than retroactively seek to ban or limit practices after they become commonplace. Just like programs that raced against the clock to field these technologies in the mid-2000s, the effort to regulate WAMI has to be a sprint operation. These things turn real far faster, and more quietly, than we might think.

---

* This lag isn't unique to drones. Law enforcement agencies have been using Sting-Ray cell phone trackers since the early 2000s, but by 2018 Congress had still not passed any national laws to regulate the technology.

The Right Kind of Principles

A cohort of advocacy groups believes wide-area surveillance ought to be banned completely. But that is neither reasonable nor practical. If the technology enables us to fight wildfires more effectively, catch hit-and-run drivers more reliably, and find disaster survivors more quickly, then we would be remiss in rejecting it whole hog.

(If anything, the cases in Dayton and Miami-Dade, where WAMI surveillance proposals were grounded by public opposition, suggest that, at least in the near future, WAMI's use may end up being naturally limited to the benign missions described in chapter 5.)

This is not to say that a community that does allow WAMI to be used specifically for law enforcement must resign itself to an unimpeded panoptic gaze, either. Law enforcement's will to reduce crime is never entirely irreconcilable with the public's desire for privacy, transparency, and accountability. The push and pull between the watchers and the watched doesn't have to be a zero-sum game.

Take the case of Santiago, Chile, where in 2015 two municipalities installed Israeli surveillance cameras mounted on balloons over two areas that had experienced a spike in muggings and assaults. Each balloon covered an area of about 30 blocks, in high resolution.

A group of local NGOs sued the government, claiming that the use of a military surveillance technology to monitor civilian populations violated the country's constitution. On a first ruling, the court of appeals of Santiago banned the balloons, but the following year, the country's supreme court ruled on appeal that the balloons could fly again, noting that the country's constitution, which was written in 1980, does not specifically guarantee a right to privacy in public space.

Nevertheless, the justices warned, the use of the technology raised concerns. "It is clear that people living under the balloons could feel watched and controlled, inducing them to change certain habits or inhibit certain behaviors within their private domestic space," the justices wrote.

The court drafted a list of conditions of use for the city government. Any surveillance activity would need to be strictly relevant to a criminal investigation. Footage that was not relevant to an investigation would have to be deleted after 30 days. The programs would be subject to monthly audits to ensure that the recordings did not capture any private activities unrelated to criminal investigations. The court also ordered the city to make the footage available to members of the public upon request.

The city government appeared to have no problem with the requirements. The mayor of one of the neighborhoods even hired an all-female team of analysts to prevent voyeurism, though this wasn't a requirement of the court.

Such a compromise could be replicated in any city hoping to use WAMI, or scaled to a national level, as long as a few simple principles are observed. First and foremost, like any stable structure, the rules and regulations need to be built on a solid foundation: a definition of the thing being regulated.

Aside from enabling stakeholders to write rules that are specific and grounded in reality, a definition is necessary to future-proof those rules against the technology's ongoing evolution, which invariably creates new loopholes for the surveiller to exploit. In the 1930s, soon after lawmakers clamped down on warrantless tapping of phone lines, engineers developed magnetic coils that could intercept calls through a wire without actually touching it, and manufacturers boasted that these devices were perfectly legal, since the standing laws only restricted physical tapping and said nothing about magnetic coils.

It would be better if we didn't have to rewrite the rules every time someone invents the WAMI equivalent of a magnetic coil — say, when a city begins using gigapixel cameras mounted on the tops of buildings or swarms of networked drones, or constellations of Cube-Sats, thus bypassing any regulations specifically written for WAMI from an airplane.

The Open Geospatial Consortium, an international organization that develops technical standards for geospatial technologies, including wide-area surveillance, defines WAMI as "a system that uses one or more *cameras* mounted on some form of a gimbal on an aircraft or blimp to capture a very large area on the ground, from about once every second up to several times per second" (emphasis in the original).

That's a little too narrow. Something like this might stick better:

Any sensor or agglomeration of sensors in the air or on the ground that generates wide, high-resolution motion imagery of the surveilled area that allows one to follow individual people and vehicles while still recording the full area.

The second crucial element for developing rules is transparency, both on the part of those who used the technology abroad in the past as well as those who intend to use it back home. As of this writing, most wide-area-surveillance programs to date, both within and beyond law enforcement, have been concealed as though they were top-secret. Nobody I spoke to would disclose which insurance companies have tested the technology for claims processing and fraud investigations, for instance, and the Forest Service even declined to share the names of the companies that manufactured the wide-area surveillance cameras it uses to fight wildfires, even though its contract announcements are public.

Such guardedness stifles public and legislative debate around how and when the technology should be used. Congress didn't debate a serious bipartisan bill to manage the use of StingRay cell-site simulators until 2017, when the devices were already widely proliferated, in

A twin-prop U-21 aircraft equipped with Blue Devil prepares to depart from Kandahar Airfield, Afghanistan, in 2011. Though it looks like any other civilian aircraft, Blue Devil is credited with assisting in the killing or capture of over 1,000 people. *The appearance of US Department of Defense (DOD) visual information does not imply or constitute DOD endorsement. US Air Force/Senior Airman David Carbajal, 451st Air Expeditionary Wing*

part because the FBI did not clarify whether it requires agencies to obtain a warrant to use them until 2015 (the 411: it usually requires a warrant, but not always). In the case of the Community Support Program, Baltimore's city council never had a chance to develop rules for WAMI because members never even knew it was being used over their heads.

A lack of transparency also exacerbates the fear factor. If you know that the police are watching you from above, but not how they're watching you, what they're watching you with, what they're looking for, or what they are and are not allowed to do with the information they collect, you might well assume that everything you do will be of interest, that even the most minor infraction could land you in the crosshairs of a military-grade surveillance apparatus.

A good start toward better transparency would be a fuller disclosure from the government as to the impact of these technologies in foreign wars. Information about the various ways in which systems like Constant Hawk, Blue Devil, and Gorgon Stare led to the capture or death of thousands of enemy combatants is entirely relevant to the debate on domestic WAMI use, as tactics and strategies honed on the battlefield can cross over into law enforcement with frightening ease. When the NYPD embarked on an aggressive — and likely illegal — surveillance campaign against a series of mosques in the mid-2000s, its tactics bore an uncomfortable resemblance to the intelligence community's "attack the network" operations in Iraq and Afghanistan, which were still, for the most part, unknown to the public. New York City's residents were never told whether "attack the network" was proving to be a successful strategy for hunting down terrorist leaders abroad, or whether it was, instead, engendering questionable wartime activities that one wouldn't want to replicate domestically.

Likewise, if WAMI has proven itself to be mostly effective for tracking and targeting large numbers of low-level combatants rather than the high-level leaders who really matter, that would suggest that when used domestically, it could be ineffective against the kingpins who keep the cogs of the drug trade in motion.

On a practical level, better transparency about past operations will cast light on the technology's other very real shortcomings. Even the technology's relatively brief history in civilian operations suggests that it doesn't always rise to the level of its formidable reputation. For instance, in 2016 the Brazilian government spent $7.5 million on four large tethered balloons equipped with Logos WAMI cameras capable of capturing 40 square kilometers at a time. The operation, which began two weeks after my visit to Baltimore, was a disaster.

Because of an anticorruption law that prohibits hardware vendors from providing operations services to government customers, Logos was barred from teaching even basic imagery analysis to the Brazilian police officers who had been assigned to operate the devices. Two of the four blimps were powered by generators, which were often left without refueling. And for reasons that remain unclear, the authorities opted to keep the blimps at 600 feet, meaning that they were only capable of surveilling a fraction of the area they were supposed to watch.

More troubling, and more telling of the technology's harmful po-

This official image of an MQ-9 Reaper equipped with Gorgon Stare undergoing a series of checks at Kandahar Airfield, Afghanistan, in 2015, prior to an operation, is probably the most detailed public image of the system ever published. Even so, it doesn't reveal much. *The appearance of US Department of Defense (DOD) visual information does not imply or constitute DOD endorsement. US Air Force/Tech. Sgt. Robert Cloys*

tential, was how, according to Logos's John Marion, the analysts spent much of their time using the cameras to watch women on the streets below. (This sort of voyeurism is not, sadly, all that unusual. In 2004 NYPD officers monitoring a protest with a surveillance helicopter specially designed for counterterrorism operations spent nearly four minutes watching a couple having sex on a balcony through the helicopter's military-type thermal camera.)

An honest conversation about what WAMI can and cannot do could help prevent similar disasters elsewhere.

Once wide-area surveillance's credentials are established, domestic agencies will want to be transparent about their intention to adopt the technology. Simply put, agencies should alert residents *prior* to acquiring WAMI — or, for that matter, any novel surveillance technology — including when the acquisition is funded by an outside source, as it was in Baltimore.

Unlike a targeted soda-straw surveillance technology that a police department can use to watch only one person at a time, WAMI watches everybody, so it is everybody's business, and every opinion counts from the start. If a police department ever adopts WAMI or any equivalent surveillance system without first submitting the proposal to the public forum, it is violating the trust of the very people it is supposed to protect and serve.

Fortunately, all local governments provide a platform for such voices to be heard. If a loud enough chorus of stakeholders opposes the use of WAMI in its particular city, as it did in Miami, the administration must not ignore it. By the same token, if enough residents support the idea of WAMI, they, too, must not be ignored. The government's job is to find a compromise. These discussions must be substantive. Relying on internet polls is no way to measure whether a community would be willing to live under wide-area surveillance.

A spirit of transparency and compromise would also help to ensure that WAMI doesn't further harm the rapport between citizens and law enforcement in places where that relationship is frayed. Since residents will be assured that the police aren't going behind their backs or over their heads — both literally and figuratively — they'll have more reason to trust their overseers, maybe even enough to cooperate with an investigation.

Finally, along with the public, one might also seriously consider bringing the technology's creators into the discussion. Nobody knows better than they what WAMI can do, how it can save lives, and how it can be abused. If the creators mean what they say about wanting the technology to have a positive impact on the world, then they need to have a chance to step up to the plate and make sure it happens — it is both their right and their responsibility.

The Right Rules

Lawmakers and their constituents. Tech firms and consumers. Engineers and privacy advocates. Any detailed regulations imposed on wide-area surveillance should be the product of a debate that involves all of these stakeholders — the rules should not come solely from a single chapter in a single book. But here are a few basic rules that could, at the very least, get this conversation started.

First, it might be wise to revisit existing legal protections from airborne intrusion and redraw the line separating "publicly accessible" technologies — which can be employed without a warrant — from "specialized technologies," which can't. There hasn't been a Supreme Court case on aerial surveillance for 30 years, and in light of the technologies that are now available, these old standards offer scant guidance as to what ought to be permissible today.

The courts could also determine whether the persistent surveillance of an individual from the sky amounts to an unreasonable search under the Fourth Amendment. There is strong legal precedent to suggest that tracking an individual over an extended period without a warrant is, in fact, unconstitutional. In two rulings in 2010 and 2012, which both stemmed from a counternarcotics operation that tried to pin the defendant to a drug network by tracking his movements with a GPS device attached to his Jeep, the DC Circuit and the Supreme Court ruled, respectively, that even though trips on public roads are plainly visible and public, the warrantless tracking of a person's journeys over an extended period (a "mosaic approach" to surveillance) constitutes a Fourth Amendment violation.

More recently, in *Carpenter v. United States*, the Supreme Court issued a similar ruling over the use of cell phone trackers for persistent surveillance. But none of these rulings specifically addressed WAMI, and so the use of WAMI can, for the moment, go on unchecked.

If the legality of wide-area surveillance can't be adjudicated through judicial process, Congress could step in. To date, there has not been a single congressional hearing that has substantially addressed the topic of WAMI, and it appears that only one Congressional Research Service report has touched on the technology's legal implications. (The study, from 2015, is primarily about drones; with respect to WAMI, it concludes, rather unhelpfully, that the use of such systems domestically raises questions about "personal control, secrecy, autonomy, and anonymity," without suggesting any WAMI-specific regulations.)

Capitol Hill could, for example, require that law enforcement agencies obtain a warrant if they wish to use WAMI to track suspects for extended periods of time, engage in counternetwork tactics that connect a wide number of actors across the surveilled area, or surveil people who are not clearly connected to a known crime, such as family members of suspects or bystanders who have witnessed an incident.

In cases where police don't have enough time to obtain a warrant — say, following an armed robbery — they could nevertheless be required to show probable cause before tracking a suspect, in much the same way that officers on the ground are required to show probable cause if they intend to follow your car, pull you over, or search your trunk.

Agencies using WAMI could also be required to set minimum operating altitudes so that individuals cannot be watched with enough resolution to be personally identified or, as with the case in Brazil, spied upon for the wrong reasons.*

To be truly watertight, these regulations would also need to ex-

---

* The higher up the plane flies, the wider the area of coverage and the lower the resolution on individuals on the ground. The lower the plane, the narrower the area it covers, but it will record individuals on the ground in much greater detail. Ross Mc-Nutt, of Persistent Surveillance Systems, says that he always favors coverage over detail. The police in Brazil evidently had the opposite view.

tend to fusion activities, given how the aggregation of data from surveillance activities that individually don't require a warrant could produce a detailed portrait of a target that is much more intrusive than the sum of its parts.

Simply following a car that made a series of suspicious U-turns might be acceptable under the probable-cause requirement, but fusing that track with a CCTV system equipped with facial recognition to identify the driver and his passengers may not. Regulations for WAMI should thus account for all the other information available to today's police forces that could be fused with it, including social media, cell phone interceptors, license plate readers, soda-straw cameras, ground cameras, and human intelligence.

Finally, these policies should, of course, be fully disclosed so that citizens solely engaged in constitutionally protected acts are reassured that the state assertively forbids surveillance of their activities.

As a way to enforce these practices, and to ensure accountability among analysts who might be tempted to, for example, spy on their spouses, surveillance and analysis activities could be meticulously logged. These logs could show whether a department watched a target persistently without a warrant. Or they could show whether an analyst's actions — say, tracking a vehicle for several hours — were related to a known crime. After a public rally, auditors could review analysis logs to make sure that the police didn't follow protesters to their homes.

In keeping with the government's own fair information practice principles, which assert that subjects of surveillance have a right to access the data that have been collected on them, police departments that use the all-seeing eye could establish a process for the public to petition for the release of logs and imagery that captured their persons or property. Following major incidents of public concern, such as police shootings, an agency could release any relevant WAMI footage in the same way that many departments have begun releasing bodycam footage in the wake of such events.

To add a further layer of protection against abuse, particularly if the data are publicly released, measures could be implemented to anonymize the surveillance imagery. This is an idea that comes straight from the WAMI industry itself. Steve Suddarth suggested that the same tracking algorithms and behavior-detection algorithms that are being developed to pursue adversary targets could also be used to mask people and vehicles that aren't directly associated with a criminal investigation.

For example, an algorithm could detect and mask every vehicle pulling out of or into a driveway or parking lot so that users who don't have a warrant are blocked from tracking individuals to a precise location (effectively preventing them from identifying the domestic equivalent of the "nodes" that the CIA uses abroad to map insurgent networks). Google already does exactly this with its cell phone tracking systems, which it uses to calculate traffic conditions on major roadways.

These are, of course, broad rules. More detailed technical norms would need to be developed, similar to the Federal Aviation Administration's strict aviation safety standards. These standards could govern details like the altitude at which the aircraft is operated, minimum and maximum camera resolutions, the length of time that data should be stored, how to track vehicles reliably, how to encrypt the airplane's data links, how the data are protected from hacking, how analysts' actions should be logged, and how publicly released imagery can be anonymized.

If these standards could be compiled into a single document or set of regulations, any new city hoping to use wide-area surveillance could adopt readymade rules, rather than having to build their own rules from scratch. If these guidelines are voluntary at the federal level, cities could adopt municipal ordinances that back them up with the force of law.

To account for those agencies that are particularly creative in their use of technology, these standards or rules could mandate that users act solely according to what *is permitted* rather than what *isn't*. For instance, if a law allows the use of surveillance to track cars and trucks but forbids the tracking of people on foot, that doesn't automatically mean that a law enforcement agency can track bicycles or scooters simply because they aren't mentioned in the rules. If an agency wishes to use wide-area surveillance in a novel way not explicitly permitted by the law, it would have to seek permission from the relevant authorities or, even better, the public at large.

Even then, because the temptation to bend or break these rules will be strong, simply trusting the authorities to adhere to them of their own volition may not be enough. A case in point: In 2015 the State of California passed a law that requires police agencies to obtain permission in a public meeting before buying a StingRay, and to publish their use policies online. But an investigation by the *Los Angeles Times* in 2017 found that almost none of the state's largest law enforcement agencies were actually following the law. Seventeen of California's 21 biggest police forces did not disclose data about their use of StingRays. Those agencies that did release records provided only sparse details about how the interceptors were used.

Similarly, as of late 2017 privacy groups in Santiago had been unable to confirm whether the Supreme Court's reforms for the surveillance balloons were being fully enforced. Though the Supreme Court had mandated a schedule of regular audits for the programs, it did not specify any guidelines to ensure that the organization that carried out the audits would be fully independent and unbiased.

Entities using wide-area surveillance could submit themselves to regular reviews by a national organization that can certify whether these rules and standards are being met, in much the same way that a variety of state and federal agencies inspect correctional facilities nationwide, often releasing the results of those audits publicly. Independent reviews of a wide-area-surveillance program could determine whether, for instance, the aircraft are being operated at an appropriate altitude and the analyst logs are being maintained and secured correctly. Any agency found to be playing fast and loose with the rules would be given a simple choice: either rectify the issue or be stripped of its eye in the sky altogether.

Private Eyes

These tactics for reining in the all-seeing eye would be geared toward public agencies — but the same gap in the law that makes it legal for the government to fly military aerial-surveillance cameras in our skies can also be exploited by any private sector company. We should therefore be just as prepared to respond to the private all-seeing eye as we are to the public one.

In some respects, keeping WAMI in private hands might have its upsides. If a company operated a wide-area surveillance system as a kind of live Google Maps accessible to everyone, from the police to traffic management authorities to investment firms, it would be more cost effective for each individual user. And it would, by design, incorporate some protections on our private data, similar to how Google blurs faces and license plates that appear in its Street View feature.

The private sector would also serve as a firewall between the government and the data hidden behind these anonymizing features. Companies like Apple provide private user data to the government only if directed to do so by a court order, and a WAMI firm could have a similar standard for revealing aerial information. Tech firms would also be in a much better position to develop technical privacy protections like the anonymizing tools discussed above.

There are also, though, serious problems with this proposition. Unlike the government, private entities will be interested in much more than just criminal activity. In a city watched over by a private wide-area-surveillance system, information about your every move could be mined and sold for a profit.

The possibilities for overstep are legion. Large companies with access to aerial-surveillance footage of an entire city could develop all kinds of clever ways of combining those data with other pieces of information to intrude upon our privacy and unveil details about activities that are supposed to be protected. Like the large social media firms, they would not always be transparent about these efforts.

And as we saw in chapter 5, once the data are made accessible to the public, even in anonymized form, it is impossible to anticipate all of the different ways that those buying the data might use them. Other forms of privately run data collection have already been used and abused in surprising and troubling ways.

In 2016 someone I'll refer to as Q was involved in a minor parking mishap that left a dent in his car. Q was very busy with work and family life at the time, and delayed filing a claim with the insurance company for several months. When Q did eventually file the report, the company denied the claim and canceled Q's policy for violating the company's terms and conditions, which require that claims be filed at the time of the accident. Somehow, it knew that Q's car was damaged long before his claim was filed. But how?

As it turned out, the company had looked up Q's street on Google View, and had spotted Q's car parked outside his house. Shortly after this happened, I, too, visited Q's address on Google Maps, and sure enough, there was the car, the dent plainly visible. Even though the license plate on Q's car had been blurred, the insurance company knew its make, model, and color, as well as Q's address (it was, in other words, doing some simple but effective spylike fusion). It had irrefutable proof, and it hadn't broken any laws.*

The rules governing private wide-area surveillance would therefore need to be just as strict and comprehensive as those we apply to government surveillance. Companies would have to be equally transparent about exactly how the data they collect are used and stored. Entities that buy this data, meanwhile, could be required to keep detailed analysis logs showing exactly what they looked at. Hopefully, Q has learned his lesson, but if his new insurer starts following the dented car around the city as though Q were an insurgent in Baghdad, Q has a right to know about it.

The Right Rules for Droids

So far, so simple. However, when *automated* surveillance comes into play, the questions that policymakers will have to answer are, as we've already seen, far knottier and unfamiliar. To recap: How do we find a balance between trusting automated systems and second-guessing them? How do we ensure that algorithms counteract bias rather than perpetuate it? How do we determine whether a system that helps police catch more criminals is worth it if it also leads to more encounters between police and innocent citizens at a time when such encounters too often end in tragedy? How do we predict and account for the bizarre—and potentially dangerous—mistakes these systems might make at any given moment?

---

* Insurance companies also sometimes use drones to spy on claimants who are suspected of lying about a work injury; in the United States, that, too, is not illegal.

Answering those questions won't be simple. But there are a few guiding principles to set us on the right course.

First, a spirit of transparency on the part of those operating the automated systems would be well-advised. As with all forms of surveillance, disclosures about what kinds of systems are in use, how those systems work, what behaviors they detect, and how they're being applied could help keep abuses in check, ease the relationship between the public and the authorities, and engender a more informed rulemaking process.

It might be worth extending demands for transparency to the companies building the software, too, so that those using it can know what to expect and those subjected to it can know what they're in for. Information of particular relevance is how the technology works, what its limitations are, and, most importantly, how it calculates its confidence level. "Never trust anything that can think for itself if you can't see where it keeps its brain," says Mr. Weasley in *Harry Potter and the Chamber of Secrets* — a statement so unintentionally true of automation that the Air Force's chief scientist even used it in formal presentations on the topic.

To further bolster that trust, more work will also be needed to make the computers themselves less mysterious and more trustworthy. One avenue for doing so is to have software essentially describe, in human language, the reasoning behind each conclusion it comes to. (DARPA's Explainable AI program intends to do just that.) With this technology, IBM's Watson might explain that it came up with the answer "Picasso" on *Jeopardy!* because Paris has a Picasso museum. Rather than losing total faith in the computer's ability to understand even simple concepts like the difference between a person and an art movement, the human operator could see that Watson was just thrown off by a small technicality. In theory, the analyst will be better prepared to know what's going through the computer's mind the next time a similar situation arises, and adjust his or her trust accordingly.

In order to try to anticipate the kinds of "Picasso" errors that these systems might make in the first place, they could be submitted to extensive testing held to common standards developed by a group like the National Institute of Justice. By extension, if an algorithm in the field ever does make a high-confidence mistake, the data that sparked the mistake could be (securely) conveyed to the manufacturer, as well as an independent oversight body, so that the algorithm can be tweaked and similar future errors can be prevented — not just for that particular department, but for every agency that uses the same software.

Another helpful measure in this regard would be standardizing confidence scoring systems among manufacturers — so that a 71 percent score from one product means the same thing as a 71 percent score from another — which could also be useful for settling some of the court cases that are likely to arise from the use of automated surveillance sooner or later.

In these inevitable cases, the judicial system would be able to establish whether a recommendation from a computer should ever be an acceptable basis for a probable-cause decision, regardless of how well it explains itself. Currently, a police officer is within his rights to pull me over if I'm driving erratically. Should the same standards apply if it was a computer that made the judgment?

When I called Chuck Blanchard, a former general counsel to both the Army and the Air Force who now works at the DC law firm of Arnold & Porter Kaye Scholer, seeking guidance on the matter, he suggested that one could use human performance in decision making as a baseline for how much or how little to trust computers. Police officers aren't always perfect logical beings, he said, and law enforcement has ways of accounting for that fallibility. He also suggested that groups might look at how trust is managed when it comes to a number of established law enforcement technologies such as breathalyzers, which are, after all, a kind of automated data-processing system.

An agency using a suspicious-driving detector might decide, for example, that it won't follow a car unless the system registers a confidence level of at least 90 percent for a given driver; with anything lower than 90 percent, the computer's findings would have to be confirmed by a pair of human eyes.

Or, more simply, one could forbid law enforcement officers from ever making probable-cause decisions purely on the basis of a computer's analysis no matter how high its confidence level, the idea being that every conclusion drawn by an autonomous system, however small, must be vetted. This would mean that systems like the Signal Innovations Group software that autonomously unravel networks by linking buildings based on whether they have the visitor patterns of restaurants or soccer fields rather than criminal nodes would be a no-go. (Some formulation of this rule might also be advantageous for managing automated intelligence in warfare.)

Another route that could mitigate some of the potential dangers of automation in law enforcement would be to ban its use in any form of proactive surveillance, limiting its applications to investigative work only. For instance, a police force could use a vehicle-tracking algorithm to map the movements of known suspects during a warranted investigation but would be prohibited from using that same software to try to anticipate crimes by detecting every car that exhibits suspicious behavior.

Limiting automation to forensic tasks would help solve the issue of creating more encounters between police and citizens, and, more significantly, it would smooth some of the trust issues that would make the use of such a system in proactive time-sensitive surveillance so dangerous. In forensic investigations, there is less time pressure, rigor is valued more than speed, and the stakes of most individual decisions are lower. There would be fewer split-second life-or-death choices made over a 71 percent confidence rate analysis.

Like the imagery itself, automated surveillance systems would also need to be secured. A hacking attack might lead a system to mark innocuous behaviors as highly suspicious — a hacker's version of a swatting call — or subtly corrupt the imagery so that it looks normal to the human eye but becomes completely confounding for the computer. If a law enforcement agency becomes too dependent on an automated system in place of human officers, such an attack that renders the system ineffective would be debilitating.

As for bias, the good news is that thanks to a number of studies, investigations, and popular books, the issue has become the subject of intense attention. So much so that even companies like Microsoft are calling for regulations that address bias in facial recognition and other tools.

When a *human* police officer displays biased behavior, law enforcement agencies have at their disposal strategies to identify that bias and, ideally, correct it through measures like training programs. If an entire agency appears to have a bias problem, the attorney general has the authority to step in to evaluate the problem, as was the case in St. Louis following the Michael Brown shooting and in Baltimore after the death of Freddie Gray.

Similarly, cities that employ automation and artificial intelligence for policing could be required to submit their programs to review by auditors who specifically search for indicators of bias in both the results generated by those systems and in the actions taken by law enforcement in response. Special care could be taken to vet the man-made data sets used for training the automated systems, since these records may have embedded biases that go back generations.

A study by a White House task force that explored the issue in 2014 even suggested that some of the very same algorithms used for analyzing data can be used to identify algorithmic bias in the results. Imagine that — computers watching computers, watching you. What a time to be alive.

One Last Pixel

Putting all of these ideas into practice may seem like a daunting challenge — but there are many signs that it is, in fact, a wholly realistic proposition. Dozens of towns and counties around the United States have already passed binding ordinances requiring law enforcement authorities to disclose the use of any new surveillance tools and comply with strict guidelines to limit abuse.

Some local governments are also already taking action on the prickly issue of automation. In 2017 New York City adopted a bill that establishes a task force to identify algorithmic bias in software used by city departments and calls for the city to set up a process by which citizens can access the data that fed an algorithmic decision by which they were personally and perhaps unjustly affected.

Meanwhile, cases like the episode in Miami, where the police department called off a plan to deploy WAMI following widespread public pushback, lend credence to the idea that the public would and could define the limits of wide-area surveillance — including, yes, barring it absolutely — and that lawmakers and law enforcement can and will respect those limits.

The stakes of getting the rules right for wide-area surveillance and automation are high. If you walk around any security trade show today, you will find a wild assortment of locked and loaded technologies on sale: autonomous robotic security guards, facial-recognition software, retinal scanners, social media trawling tools, computer-vision systems for bodycams, and a staggering number of fusion systems. All of these technologies operate on the same underlying formula — to get left of the crime, collect as much information as possible, and automate the analysis. Many of the systems have, like WAMI, a secret violent past in foreign battlefields. All could be abused. None has been regulated. Coming up with a workable model for wide-area surveillance will prove that these technologies can be held on a straight line, too.

We should also recognize, however, that rules can be a bit of a poisoned chalice. They have the potential to legitimize the technology and, as a result, hasten its proliferation. Shortly after the Chilean Supreme Court ruling on the continued use of a fleet of surveillance balloons over Santiago, the city's newly elected mayor announced a plan to install additional balloons over three of the city's "most dangerous" neighborhoods. It was as though the ruling had fully put to bed the question of whether the balloons could ever again represent a credible threat to privacy; in fact, it hadn't.

Many questions remained, and in one way or another they always will. Even regulations that seem completely watertight at first will begin taking on water sooner or later. As the technology proliferates, new, as-yet-unthought-of uses for the all-seeing eye will create new quandaries. Newly developed capabilities will provoke entirely new questions. Even if the engineers and commanders behind those new iterations of the technology have the best intentions, the groups who buy them in the back rooms of a future trade show may not.

This is where those of us who have been kept out of the first chapters of the wide-area surveillance story finally come in. And our job is clear. We need to demand transparency from our own overwatchers and those who overwatch others. We have to make sure the rules stay ahead of the tools. In making those rules, we will fare best if we work with, not against, those on the other end of the camera—whether they be the engineers, commanders, enforcers, or advocates.

And remember, the surveiller's quest is a long road. As such, the job of responding to surveillance, of making sure it's done right, is also necessarily a never-ending one. So stay on the case. And always keep an eye on the sky.

Acknowledgments

First, I must reserve special thanks for Dan Gettinger, my colleague, fellow soft-serve aficionado, and friend. Dan was kind enough to deploy his astounding research powers to help me navigate a set of particularly obscure DOD budget documents for this book. More to the point, were it not for Dan, the Center for the Study of the Drone would never have succeeded — for this and for so much more, I am greatly indebted to him. I am also grateful for the entire Bard College family, particularly Tom Keenan and Jim Brudvig, the development and alumni team, and Leon Botstein — all took a chance on the Center for the Study of the Drone that no one but I, as a scrappy undergraduate, asked them to take.

Talking to journalists does not come naturally to people in the world of spycraft, and though over the course of reporting this book I was declined answers to hundreds of questions that apparently prodded a little too probingly at the thick black veil of government secrecy, not a single source ever questioned my right to ask those questions in the first place. I would particularly like to thank Steve Suddarth, who let me fly with him in his surveillance plane over Albuquerque and then, the following day, chased me around the New Mexico desert with that very same aircraft; K. Palaniappan and his team for processing the footage from those flights; and Ross McNutt for bringing me into what was at the time his secret surveillance operation in Baltimore; as well as John Marion, Nathan Crawford, and Brian Leininger for bearing with me through extensive fact-checking discussions and for providing a number of the excellent images found throughout this text.

On the other side of the page, I am profoundly grateful to Eamon

Dolan, my original editor—and one of grand caliber, at that—my second editor, Alex Littlefield, another maven who carried the book across the final mile just as he became a father to lovely baby Frieda; and Olivia Bartz, a rising star who, in turn, stepped up when Alex was called away for fathering duties. Their dedication to the project meant a great deal to me. Of my many mentors and guardian angels in the arena, I owe special credit to Shaye Areheart for writing the email, in November 2015, that set me off on this whole adventure.

The recipient of that email was the brilliant Howard Morhaim, a man of exacting standards, exquisite taste, and superlative wisdom, who, in spite of all these traits, agreed to become my agent. I am very glad he did. Not only is Howard the best in the business; he has become, much more importantly, a truly wonderful friend.

Notes

Introduction

page

xi   *"We watched with anxious eyes"*: William Alexander Glassford, "The Balloon in the Civil War," *Journal of the Military Service Institution of the United States* 18, no. 80 (1896): 259–60.

xiii *"many times deadlier"*: W. Benjamin, "The Camera at the Front," *Scientific American,* November 24, 1917.

xiv *it was a philosophy that begat:* Ellen Nakashima and Joby Warrick, "For NSA Chief, Terrorist Threat Drives Passion to 'Collect It All,'" *Washington Post,* July 14, 2013, https://www.washingtonpost.com/world/national-security/for-nsa-chief-terrorist-threat-drives-passion-to-collect-it-all/2013/07/14/3d26ef80-ea49-11e2-a301-ea5a8116d211_story.html?utm_term=.d9cbfe7c5853.

1. A New Threat

3   *Matthew Bourgeois was killed:* "HMC (SEAL) Matthew Bourgeois," Pritzker Military Museum & Library, www.pritzkermilitary.org/explore/museum/past-exhibits/seal-unspoken-sacrifice/matthew-bourgeois/.
four US soldiers died: "Car Bomb Near Najaf Kills 4 US Soldiers," United Press International, March 29, 2003, www.upi.com/Defense-News/2003/03/29/Car-bomb-near-Najaf-kills-4-US-soldiers/74231048931690/.
when an object exploded: "Army Pfc. Jeremiah D. Smith," *Honor the Fallen, Military Times,* thefallen.militarytimes.com/army-pfc-jeremiah-d-smith/256671.
"unexploded ordnance": Ibid.
wrote a classified memo: Andrew Smith, *Improvised Explosive Devices in Iraq, 2003–09: A Case of Operational Surprise and Institutional Response* (Carlisle, PA: US Army War College, Strategic Studies Institute, April 2011), 13.

4   *identified their targets:* Steve Edgar, interview with the author, May 26, 2016.
several months old: Ibid.
"unparalleled in the history of air warfare": Harold P. Myers, *Nighthawks over Iraq: A Chronology of the F-117A Stealth Fighter in Operations Desert Shield and*

Desert Storm, Special Study 37FW/HO-91-1 (Langley, VA: Office of History Headquarters, 37th Fighter Wing Twelfth Air Force Tactical Air Command, January 9, 1992), 2–4.

5   *all anyone needed to escape the US Air Force:* James Poss, interview with the author, July 7, 2016.

Donald Rumsfeld called for the Pentagon: James McCarthy et al., *Transformation Study Report, Executive Summary: Transforming Military Operational Capabilities* (Washington, DC: Department of Defense, April 27, 2001), 10.

had shown that it was easy to shoot down: James Risen and Ralph Vartabedian, "Spy Plane Woes Create Bosnia Intelligence Gap," *Los Angeles Times,* December 2, 1995, http://articles.latimes.com/1995-12-02/news/mn-9494_1_military -intelligence.

Chances were that a Predator's quarry: Poss, interview, July 7, 2016.

6   *within a hair of turning the course of history:* Arthur Holland Michel, "How Rogue Techies Armed the Predator, Almost Stopped 9/11, and Accidentally Invented Remote War," *Wired,* December 17, 2015, www.wired.com/2015/12/ how-rogue-techies-armed-the-predator-almost-stopped-911-and-accidentally -invented-remote-war/.

7   *an operation in 2004 called "the Blitz":* Rick Atkinson, "Left of Boom: 'There Was a Two-Year Learning Curve . . . and a Lot of People Died in Those Two Years,'" *Washington Post,* October 1, 2007.

"your job": Brad Ward, interview with the author, June 16, 2016.

8   *"Oh," they'd say. "We missed that":* Ibid.

instead of trying to find: Zachary Lemnios, "Priority Briefing—Zachary Lemnios, CTO, Lincoln Laboratory," *Lincoln Laboratory Journal,* May 2009, www.milsatmagazine.com/story.php?number=1920828481.

a new doctrine of airborne intelligence emerged: United States Joint Forces Command, *Commander's Handbook for Persistent Surveillance,* Version 1.0 (Suffolk, VA: Joint Warfighting Center, Joint Doctrine Support Division, June 20, 2011), 6.

"the eyes don't go off the target": Poss, interview, July 7, 2016.

spent 630 hours watching Abu Musab al-Zarqawi: Dan Murphy and Mark Sappenfield, "A Long Trail to Finding Zarqawi," *Christian Science Monitor,* June 9, 2006.

you learned everything about him: Poss, interview, July 7, 2016.

a rare interview near the end of his tenure: "Interview: Gen. Michael Wooley, Air Commandos' Unblinking Eye," IntelliBriefs, April 2, 2007, http://intelli briefs.blogspot.com/2007/04/interview-gen-michael-wooley-air.html.

"the God's eye view": Ward, interview, June 16, 2016.

9   *Ward didn't like this:* Ibid.

They were like "spear fishermen": Michael Flynn, *Charlie Rose,* PBS, February 4, 2015, accessed via Internet Archive, https://archive.org/details/ KQED_20150204_200000_Charlie_Rose/start/660/end/720.

a morbid game of whack-a-mole: "IEDs: The Home-Made Bombs That Changed Modern War," *Strategic Comments* 18, no. 5 (2012): 1–4.

"What we really needed": Flynn interview, *Charlie Rose.*

10   *A leaked 2011 Scientific Advisory Board study concluded:* United States Air Force Scientific Advisory Board, *Operating Next-Generation Remotely Piloted Aircraft for Irregular Warfare,* SAB-TR-10-03 (Washington, DC: US Airforce Scientific Advisory Board, April 2011), 6.

11   *"part art, part luck":* Scott Swanson, interview with the author, April 29, 2016. *Even following a single vehicle:* John Montgomery, interview with the author, December 19, 2016.

Predators would keep their cameras tightly zoomed in on their targets: Mark Cooter, interview with the author, May 31, 2017.

a single Predator assigned to watch over a 40-square-mile area: Michael W. Isherwood, *Layering ISR Forces,* Mitchell Paper 8 (Arlington, VA: Mitchell Institute Press, 2011), 15.

The Predator crews watching Zarqawi: Murphy and Sappenfield, "A Long Trail to Finding Zarqawi."

12   *closed the door to the office and produced a classified document:* Steve Suddarth, interview with the author, September 20, 2016.

a problem that matched the "complexity and urgency": Smith, *Improvised Explosive Devices in Iraq, 2003–09,* 13.

the senior command was willing to try anything: Atkinson, "Left of Boom: 'There Was a Two-Year Learning Curve.'"

had taken to tying leaf blowers to the bumpers of their Humvees: Ibid.

Cabayan believed that if any organization: Suddarth, interview, September 20, 2016.

13   *One of the more infamous products to emerge from the project:* Noah Shachtman, "Military Security Threat: Bogus Bomb-Zapper's Bogus Countermeasure," *Wired,* July 16, 2007, https://www.wired.com/2007/07/nobody-wants-re/.

would go on to spend over $75 billion: Gregg Zoroya, "How the IED Changed the US Military," *USA Today,* December 18, 2013, https://www.usatoday.com/story/news/nation/2013/12/18/ied-10-years-blast-wounds-amputations/3803017/.

laser-induced breakdown spectroscopy systems: Daniel Díaz, David W. Hahn, and Alejandro Molina, "Laser-Induced Breakdown Spectroscopy (LIBS) for Detection of Ammonium Nitrate in Soils," *SPIE Proceedings* 7303, Detection and Sensing of Mines, Explosive Objects, and Obscured Objects 14 (2009), doi: 10.1117/12.818391.

specialized drones; ground-penetrating radars; electromagnetic detection systems: Clay Wilson, *Improvised Explosive Devices (IEDs) in Iraq: Effects and Countermeasures* (Washington, DC: Congressional Research Service, February 10, 2006), 4.

he refused to discuss any of them in detail: Suddarth, interview, September 20, 2016.

2. Enemy of the State

14 *One Friday evening in the winter of 1998:* This account was provided by the media relations office at Lawrence Livermore, which acted as intermediary between the author and the individuals involved.

called it the NSA's "Big Daddy": Tony Scott in *The Making of "Enemy of the State."* DVD, Buena Vista Home Entertainment, 2006.

15 *Lawrence Livermore was . . . US airspace:* Bart Hacker, "A Short History of the Laboratory at Livermore," *Science & Technology Review,* September 1998.

16 *at an altitude of 300 kilometers:* Readers will note that both US and metric units of measure appear herein. Since the metric system prevails in science and engineering, both in the US and globally, the author has opted to use that system in those contexts. In most nontechnical contexts, and where aesthetic considerations dictate, US measure is given.

had proposed a project called Brilliant Eyes: Edward Teller, "Brilliant Eyes," in *International Seminar on Nuclear War, 10th Session: Planetary Emergencies,* Science and Culture Series: Nuclear Strategy and Technology (Singapore: World Scientific, 1992), 20.

17 *their thoughts first turned to nonproliferation operations:* John Marion, interview with the author, June 7, 2017.

Saddam Hussein was mostly refusing to cooperate: "Iraq: A Chronology of UN Inspections," Arms Control Association, https://www.armscontrol.org/act/2002_10/iraqspecialoct02.

colleagues from another department at Livermore: Walter Pincus, "N. Korea's Nuclear Plans Were No Secret," *Washington Post,* February 1, 2003, https://www.washingtonpost.com/archive/politics/2003/02/01/n-koreas-nuclear-plans-were-no-secret/cdc1f774-a857-4732-a1cb-86fc78637d82/?noredirect=on&utm_term=.9e945c0038b2.

For its part, the CIA had begun to collect reports: Director of National Intelligence, *Unclassified Report to Congress on the Acquisition of Technology Relating to Weapons of Mass Destruction and Advanced Conventional Munitions, 1 January Through 30 June 2002* (Washington, DC: Central Intelligence Agency, October 9, 2002).

"If you have an area where you know bad stuff is going to happen": Marion, interview, June 7, 2017.

Marion pitched the idea: John Marion, email to the author, April 30, 2018.

18 *that system was capable of capturing a wide area:* Ibid.

A German company . . . Livermore's home-brew unit cost just $80,000: D. M. Pennington, *Sonoma Persistent Surveillance System,* UCRL-TR-220175 (Livermore, CA: Lawrence Livermore National Laboratory, March 28, 2006).

19 *To replicate . . . deeply impressed:* Marion, interview, June 7, 2017.

the rare ability to film a golf ball: Ibid.

The new, improved camera first flew: Nathan Crawford, interview with the author, January 26, 2017.

In one flight simulating the arc of a satellite in orbit: Ibid.

20 *at least eleven Nobel laureates:* Ann Finkbeiner, *The Jasons: The Secret History of Science's Postwar Elite* (New York: Viking, 2006), 4.

concluded that the use of small nuclear weapons: F. J. Dyson et al., *Tactical Nuclear Weapons in Southeast Asia,* Study S-266 (Alexandria, VA: Institute for Defense Analyses, Jason Division, March 1967).

One early Jason study . . . advanced remote-sensing techniques: John Horgan, "Rent-a-Genius," *New York Times,* April 16, 2006, www.nytimes.com/2006/04/16/books/review/rentagenius.html.

More recently, the group has weighed in on the merits: Federation of American Scientists, "JASON Defense Advisory Panel Reports," updated February 1, 2018, https://fas.org/irp/agency/dod/jason/.

In the fall of 2003: Dan Cress, email to the author, May 16, 2018.

an expert in seismic, acoustic, and electromagnetic surveillance: Office of the Director of National Intelligence, Public Affairs Office, "Director of National Intelligence Fellows Award Winners Announced," ODNI News Release No. 24-06, December 15, 2006, https://www.dni.gov/files/documents/Newsroom/Press%20Releases/2006%20Press%20Releases/20061215_release.pdf.

In the Jasons' first meeting: Cress, email, May 16, 2018.

he didn't see it as the CIA's job: Dan Cress, interview with the author, October 26, 2016.

21 *In a single 12-month span:* John Marion, interview with the author, February 10, 2017.

Marion had added a few slides to his presentation: Marion, interview, June 7, 2017.

he invited Marion: Cress, email, May 16, 2018.

There were multiple disparate IED cells: Blake Morrison and Peter Eisler, "Commanders Pushed to Make Bomb Disposal Choices," *USA Today,* November 14, 2007, usatoday30.usatoday.com/news/military/2007-11-06-eod_N.htm.

22 *Marion was the next speaker:* Marion, interview, February 10, 2017.

Predator crews already employed this strategy: Ward, interview, June 16, 2016.

23 *each IED cell consisted:* Wilson, *Improvised Explosive Devices (IEDs) in Iraq,* 2.

Each vehicle . . . "node": United States Joint Forces Command, *Commander's Handbook for Attack the Network,* Version 1.0 (Suffolk, VA: Joint Warfighting Center, Joint Doctrine Support Division, May 20, 2011), III-3.

To prove they could follow vehicles in this way: Marion, interview, June 7, 2017.

if you're dealing with a true network: Keith Masback, interview with the author, June 30, 2016.

One proxy might travel between two known significant nodes: Interview with Dan Cress, October 26, 2016.

With a full picture of the group's structure . . . disrupting its operations: US Joint Forces Command, *Commander's Handbook for Attack the Network*, Version 1.0, IV-3.

24 *he will never forget Marion's briefing:* Roy Schwitters, email to the author, May 16, 2018.

The group decided to immediately draft a classified two-page letter: Cress, interview, October 26, 2016.

Cress arranged a meeting with Phillips and Riley . . . agency and the Pentagon: Ibid., and Cress email to author, May 16, 2018.

"the Godfather": Marion, interview, June 7, 2017.

Crawford flew a second prototype camera to San Diego: Marion, interview, June 7, 2017.

the engineers had built computer software: William Ross, interview with the author, December 14, 2016.

25 *anything that was actually moving:* Crawford, interview, January 26, 2017.

the most amazing thing he had ever seen: Suddarth, interview, September 20, 2016.

Suddarth was so impressed . . . system of their own: Suddarth, interview, September 20, 2016.

Suddarth felt that Livermore's prototype has a crucial drawback . . . they worked through the battlefield: Suddarth, interview, September 20, 2016.

26 *When Dan Cress heard about Suddarth's idea:* Cress, interview, October 26, 2016, and email, May 16, 2018.

27 *the researchers in New Mexico opted to build their own hardware:* Suddarth, interview, September 20, 2016.

In a meeting in a secured room . . . "a picture of the driver": Steve Suddarth, interview with the author, June 9, 2017.

Thanks to the Jasons' weighty endorsement . . . development, testing, and preparations for deployment: According to various discussions and timelines provided to the author.

28 *Suddarth brought with him . . . wanted to impress upon them:* Suddarth, interview, June 9, 2017, and Ross McNutt, interview with the author, June 13, 2016.

would receive funding spasmodically: Bill Hoffman, interview with the author, January 17, 2017.

He built large, well-appointed trailers for the engineers: Crawford, interview, January 26, 2017.

29 *It was common for engineers on board to develop symptoms of oxygen deprivation:* Marion, interview, June 7, 2017.

a single bombing in Baghdad killed 112 people: Rory Carroll, "Iraq Bombings and Shootings Leave 150 Dead," *Guardian*, September 15, 2005, https://www.theguardian.com/world/2005/sep/15/iraq.rorycarroll.

"There started to be a figurative body count on this program": Crawford, interview, January 26, 2017.

the team had flown the camera over Otay Mesa: Marion, interview, June 7, 2017, and John Marion, email to the author, August 8, 2018.

though Crawford understood the project's potential: Crawford, interview, January 26, 2017.

30 *When I first spoke to Crawford on the phone:* Ibid.

With their 66 million pixels: James A. Ratches, Richard Chait, and John W. Lyons, *Some Recent Sensor-Related Army Critical Technology Events* (Washington, DC: National Defense University, Center for Technology and National Security Policy, February 2013), 10, and Marion, interview, June 7, 2017.

the system could watch: Pennington, *Sonoma Persistent Surveillance System*, 19.

In order to turn the video: Henry Canaday, "Unmanned Eyes," Intelligence Geospatial Forum, May 8, 2014, http://www.kmimediagroup.com/gif/articles/434-articles-cgf/unmanned-eyes.

31 *it would have made that year's TOP500:* William Ross, interview with the author, December 14, 2016.

In a final test flight in February 2006: Based on various accounts from John Marion and Nathan Crawford in 2017 and 2018.

The local papers and the subsequent accident investigation: "Coroner Says Plano Man Among Three Killed in Plane Crash," *My Plainview*, February 5, 2006, https://www.myplainview.com/news/article/Coroner-says-Plano-man-among-three-killed-in-8689835.php.

five airplanes were dispatched: Ratches, Chait, and Lyons, *Some Recent Sensor-Related Army Critical Technology Events*, 4.

US forces in Iraq were encountering: Robert Bryce, "Surge of Danger for US Troops," *Salon*, January 22, 2007, www.salon.com/2007/01/22/ieds/.

Numerous Pentagon officials told Crawford: Crawford, interview, January 26, 2017.

The Army also began working: Ratches, Chait, and Lyons, *Some Recent Sensor-Related Army Critical Technology Events*, 9.

32 *The following year:* Spencer Ackerman, "High-Tech Army Team Turns from Killers to Airborne Spies," *Wired*, August 16, 2010, www.wired.com/2010/08/high-tech-army-task-force-turns-from-killers-to-airborne-spies/; and "Night Eyes for the Constant Hawk—Opening the Night for Counter-IED Surveillance," Defense Update, September 17, 2009, defense-update.com/20090917_awapss.html.

Whereas a single Predator assigned to a 40-square-mile area: Isherwood, *Layering ISR Forces*, 15.

33 *Two of the aircraft were equipped:* Ratches, Chait, and Lyons, *Some Recent Sensor-Related Army Critical Technology Events*, 10.

couriers would personally deliver: Marion, email, April 30, 2018.

and an NGA facility in Virginia: Khoi Nguyen, "Aerial ISR Processing, Exploitation, Dissemination (PED)," PowerPoint presentation, US Army, PM Sensors–Aerial Intelligence, March 12, 2014.

Cress would later only say: Cress, interview, October 26, 2016.

a site referred to by one Army document: Nguyen, "Aerial ISR Processing, Exploitation, Dissemination (PED)."

one defense executive: Edwin C. Tse, "IMSC Spring Retreat Activity Based Intelligence Challenges," PowerPoint presentation, Integrated Media Systems Center Spring Retreat, University of Southern California, Davidson Conference Center, March 7, 2013.

zip ties and "good intentions": Charles Law, interview with the author, January 17, 2017.

A single variant of the system: BAE Systems, "High-Resolution Airborne Surveillance Sensor Surpasses 10,000 Flight Hours," press release, May 1, 2013, www.baesystems.com/en-us/article/highresolution-airborne-surveillance sensor-surpasses-10000-flight-hours.

34 *the Pentagon established multiple intelligence cells:* Marion, email, April 30, 2018, and LinkedIn profile of Edinoel Soto, an intelligence analyst who describes "Set[ting] up new CH analyst sites at multiple locations in the DC Metro area": https://www.linkedin.com/in/edinoel-soto-44449110/.

he would only say that the program very quickly succeeded: Crawford, interview, January 26, 2017.

Crawford believes that up to 600 US service members: Crawford, interview, February 7, 2017.

Task Force ODIN is said to have "eliminated": Kris Osborn, "Army Sends ODIN to Afghanistan," *Army Times*, December 15, 2008, posted on Tapatalk. com, https://www.tapatalk.com/groups/warships1discussionboards/us-army odin-unit-heads-to-afghanistan-t8092.html.

3. The Gorgon's Stare

35 *had requested Angel Fire after watching it in action:* Tom Vanden Brook, "Spy Technology Caught in Military Turf Battle," *USA Today*, October 17, 2007, usatoday30.usatoday.com/news/military/2007-10-02-angel-fire_N. htm.

All told, the aircraft flew more than 1,000 sorties: Daniel Alan Uppenkamp, "Two Fundamental Building Blocks to Provide Quick Reaction Capabilities for the Department of Defense" (MS dissertation, Department of Computer Science, Wright State University, 2013), 38, 55.

when analysts discovered a location: Ross McNutt, interview with the author, June 13, 2016.

36 *McNutt—whose technical contributions, Suddarth claims:* Suddarth, interviews, September 20, 2016, and June 10, 2017.

McNutt was frustrated: McNutt, interview, July 13, 2016.

The Marine Corps accused various Army officers... their camera over the Army's: Vanden Brook, "Spy Technology Caught in Military Turf Battle."

That same year: Ibid.

Senator Kit Bond, a Republican from Missouri: Ibid.

37 *In 2014 he was inducted to the shadowy:* Freedom Through Vigilance Association, *Air Force Intelligence, Surveillance and Reconnaissance Agency Hall of Honor Induction & Anniversary Banquet,* pamphlet, September 27, 2014.

in 1996 he had convinced his boss: Richard Whittle, *Predator: The Secret Origins of the Drone Revolution* (New York: Holt, 2014).

Meermans noticed... might be possible: Michael Meermans, interview with the author, August 2, 2016.

When Meermans and his colleagues... "we're going to build one": Ibid.

The new camera: US Senate, 110th Cong., 2nd Sess., Report 110-225, *Report Authorizing Appropriation for Fiscal Year 2009 for Military Activities of the Department of Defense, for Military Construction, and for Defense Activities of the Department of Energy, to Prescribe Personnel, Strengths for Such Fiscal Year, and for Other Purposes,* May 12, 2008, https://www.congress.gov/congression al-report/110th-congress/senate-report/335/1.

And, unlike Constant Hawk and Angel Fire: Meermans, interview, August 2, 2016, and August 5, 2016.

38 *combat TiVO!:* Poss, interview, July 7, 2016.

Later, a military funding bill called the performance targets: US Senate, 11th Cong., 1st Sess., Report 111-35, *National Defense Authorization Act for Fiscal Year 2010: Report Authorizing Appropriations for Fiscal Year 2010 for Military Activities of the Department of Defense,* July 2, 2009, https://www.gpo.gov/fdsys/pkg/CRPT-111srpt35/html/CRPT-111srpt35.htm.

suicide bombers had killed 50 Iraqi civilians: "Baghdad Bombers Target Soccer Celebrations, Killing at Least 50," CNN, July 25, 2007, www.cnn.com/2007/WORLD/meast/07/25/iraq.main/index.html.

39 *The contractor designed the remote-operations system over the course of a few weeks:* Arthur Holland Michel, "How Rogue Techies Armed the Predator, Almost Stopped 9/11, and Accidentally Invented Remote War," *Wired,* December 17, 2015, https://www.wired.com/2015/12/how-rogue-techies-armed-the-predator-almost-stopped-911-and-accidentally-invented-remote-war/.

40 *"Those who say it cannot be done should not get in the way of those doing it":* Interview with Michael Meermans, August 5, 2016.

On his first day at the company: Meermans, interview, August 2, 2016.

So someone in the community: Ibid.

"rigid, fixed, penetrating, unblinking stare": Stephen R. Wilk, *Medusa: Solving the Mystery of the Gorgon* (New York: Oxford University Press, 2000), 124.

the Gorgon is described as having "burning eyes": Homer, *The Iliad,* trans. Robert Fagles (London: Penguin, 1990), bk. 11, line 40.

41 *"dreadful Queen Persephone might send":* Homer, *The Odyssey,* trans. Emily Wilson (New York: Norton, 2017), bk. 11, lines 634–35.

"Ultimately . . . it made a lot of sense": Meermans, interview, August 2, 2016.

they liken to picking athletes for a sports team: Ibid.

hypersonic cruise missiles: Mark Gustafson, "Hypersonic Air-breathing Weapon Concept (HAWC)," US Defense Advanced Research Projects Agency, www.darpa.mil/program/hypersonic-air-breathing-weapon-concept.

programmable microorganisms: US Defense Advanced Research Projects Agency, "Biological Technologies Office (BTO)," www.darpa.mil/about-us/offices/bto.

42 *When Brian Leininger set out:* Brian Leininger, interview, February 6, 2017.

Leininger decided to approach Tether: Ibid.

also believed that the cameras ought to cover a much wider area: Ibid.

Fifty percent of the vehicles of interest: William Ross, interview with the author, December 14, 2016.

And because people on foot move in less predictable ways: Leininger, interview, February 6, 2017.

Tether had resisted their entreaties: Marion, interview, June 7, 2017.

43 *Leininger had also decided that the system needed to be small:* Richard Nichols, interview with the author, February 24, 2017.

Nichols thought he was "off his rocker": Ibid.

A camera on the order that he was proposing would be a technological nightmare: Marion, interview, February 10, 2017.

a faster frame rate, in turn, requires faster computers: Dwayne Jackson, David Lamartin, and Jacqueline Yahn, *WAMI Final Report* (Washington, DC: Secretary of the Air Force for Acquisition, December 11, 2012), 20.

DARPA approved the full proposal in November 2006: Nichols, interview, February 24, 2017.

Argus is a fearsome giant: "Argus," Encyclopaedia Britannica Online, https://www.britannica.com/topic/Argus-Greek-mythology.

44 *"The herdsman Argus":* Elizabeth Barrett Browning, "Prometheus Bound," in *Prometheus Bound and Other Poems; Including Sonnets from the Portuguese, Casa Guidi Windows, etc.* (New York: C. S. Francis, 1851), 36.

they would only believe the technology was possible: Nichols, interview, February 24, 2017.

just shook their heads: Leininger, interview, February 6, 2017.

simply strapping more individual video cameras together: William Ross, interview with the author, December 14, 2016.

Using a five-millimeter-wide cell phone camera chip: Ibid.

45 *It occurred to Ross and his colleagues:* Ross, interview, February 2, 2018.

With funding from the Rapid Reaction Technology Office: Ross, interview, December 14, 2016.

arranged 176 of the five-millimeter cell phone camera chips into four identical grids: Ibid., and Richard Sinn, "Virtual Pan-Tilt-Zoom for a Wide-Area-Video Surveillance System" (master's thesis, Department of Electrical Engineering and Computer Science, MIT, August 22, 2008), 21.

DARPA awarded the British defense firm BAE Systems: Yiannis Antoniades, interview with the author, February 16, 2017.

46   *signed a memorandum of agreement:* Richard Nichols, email to the author, March 11, 2017.

selected cell phone chips made by a company called Micron: Antoniades, interview, February 16, 2017.

BAE proposed to stitch the chips: Brian Leininger et al., "Autonomous Real-Time Ground Ubiquitous Surveillance-Imaging System (ARGUS-IS)," *SPIE Proceedings* 6981, Defense Transformation and Net-Centric Systems (2008), doi: 10.1117/12.784724, p. 5.

47   *was doubling the capacity of its GPUs:* Wired Staff, "NVIDIA," *Wired*, July 1, 2002, https://www.wired.com/2002/07/nvidia/.

a researcher named Sheila Vaidya, who had been leading: Ann Parker, "Built for Speed: Graphics Processors for General-Purpose Computing," *Science & Technology Review*, November 2005.

a similar unit using the insides: Ross, email to the author, May 8, 2018.

these computers were much more powerful: Arnie Heller, "From Video to Knowledge," *Science & Technology Review*, April/May 2011.

packed with 33,000 processing elements: Brian Leininger, "Autonomous Real-Time Ubiquitous Surveillance-Imaging System (ARGUS-IS)," PowerPoint presentation, The Villages Science and Technology Club Meeting, The Villages, Florida, February 8, 2016.

so much processing power been packaged Antoniades, interview, February 16, 2017.

the engineers looked at a neighboring facility's car park: Nichols, interview, February 24, 2017.

48   *ARGUS had 1,854,296,064 pixels:* Leininger et al., "Autonomous Real-Time Ground Ubiquitous Surveillance-Imaging System (ARGUS-IS)," 1.

It generated 27.8 gigabytes of raw pixel data: National Research Council, Committee on Developments in Detector Technologies; Standing Committee on Technology Insight—Gauge, Evaluate, and Review; Division on Engineering and Physical Sciences, *Seeing Photons: Progress and Limits of Visible and Infrared Sensor Arrays* (Washington, DC: National Academies Press, 2010), 126. Calculation based on ARGUS-IS's 424-gigabit-per-second data rate and a 25mbps wireless connection, which is roughly equivalent to the fastest services provided in the United States in 2017 according to Kevin

Murnane, "Speedtest's 2017 Report Ranks the Fastest Mobile Carriers in the US," *Forbes*, September 12, 2017, https://www.forbes.com/sites/kevinmur nane/2017/09/12/speedtests-2017-report-ranks-the-fastest-mobile-carriers-in-the-us/#3a7efac7651a.

Just processing all the pixels: Leininger et al., "Autonomous Real-Time Ubiquitous Surveillance-Imaging System (ARGUS-IS)."

With the Black Hawk hovering . . . windshield wipers on the cars: Leininger, "Autonomous Real-Time Ubiquitous Surveillance-Imaging System (ARGUS-IS)."

"I think we've done something good here": Antoniades, interview, February 16, 2017.

50   *On the day the demonstration was scheduled, poor weather prevented the agency from putting on:* Nichols, interview, February 24, 2017.

there were more than 7,000 IED attacks: Rob Evans, "Afghanistan War Logs: How the IED Became Taliban's Weapon of Choice," *Guardian*, July 25, 2010, https://www.theguardian.com/world/2010/jul/25/ieds-improvised-explosive-device-deaths.

Big Safari did not want to wait for DARPA: Ed Topps, interview with the author, June 30, 2017.

considered it to be one of his highest priorities: Ibid.

"GET THAT INDIVIDUAL": Topps, email to the author, June 28, 2017.

the Marine Corps and the Air Force had even canceled an order: US Senate, 111th Cong., 1st Sess., Report 111-35, *National Defense Authorization Act for Fiscal Year 2010: Report Authorizing Appropriations for Fiscal Year 2010 for Military Activities of the Department of Defense* (Washington, DC: US Government Printing Office, 2009), https://www.gpo.gov/fdsys/pkg/CRPT-111srpt35/html/CRPT-111srpt35.htm.

which had been developed by ITT Corporation: US House of Representatives, 110th Cong., 1st Sess., Report 110-477, *National Defense Authorization Act for Fiscal Year 2008: Conference Report to Accompany H.R. 1585* (Washington, DC: US Government Printing Office), 769, https://www.gpo.gov/fdsys/pkg/CRPT-110hrpt477/html/CRPT-110hrpt477.htm.

roughly the same pixel count as Livermore's prototype: Jackson, Lamartin, and Yahn, *WAMI Final Report*, 7.

The cameras and their accompanying processors: Ibid.

51   *the Air Force was insisting . . . be terminated instead:* US Senate, 111th Cong., 1st Sess., Report 111-35, *National Defense Authorization Act for Fiscal Year 2010: Report Authorizing Appropriations for Fiscal Year 2010 for Military Activities of the Department of Defense* (Washington, DC: US Government Printing Office, 2009), https://www.gpo.gov/fdsys/pkg/CRPT-111srpt35/html/CRPT-111srpt35.htm.

It could only transmit . . . failures per flight: Memorandum, "MQ-9 Gorgon

Stare (GS) Fielding Recommendation," US Air Force, Air Combat Command, 53rd Wing, December 30, 2010.

"You're speeding": Topps, interview, June 30, 2017.

"DO NOT field": Memorandum, "MQ-9 Gorgon Stare (GS) Fielding Recommendation."

An Air Force spokesperson issued a response: Lt. Col. Richard Johnson, "Statement on Gorgon Stare," January 25, 2011, available via website of *Air Force Magazine,* http://www.airforcemag.com/SiteCollectionDocuments/Reports/2011/January%202011/Day25/GorgonStare_012511.pdf.

Air Combat Command was threatening: Topps, interview, June 30, 2017, and Poss, interview, July 7, 2016.

more than $500 million: Yochi J. Dreazen, "Internal Air Force Report: New Drone 'Not Operationally Effective," NextGov, January 25, 2011, www.nextgov.com/defense/2011/01/internal-air-force-report-new-drone-not-oper ationally-effective/48362/.

Speaking to reporters from the Washington Post*:* Ellen Nakashima and Craig Whitlock, "With Air Force's Gorgon Drone 'We Can See Everything,'" *Washington Post,* January 2, 2011, www.washingtonpost.com/wp-dyn/content/arti cle/2011/01/01/AR2011010102690.html.

When I asked Poss about the article in 2016: Poss, interview, July 7, 2016.

52 *Topps showed Cartwright the live surveillance footage:* Topps, interview, June 30, 2017.

The first four Gorgon Stare–equipped Reaper drones: US House of Representatives, 113th Cong., 1st Sess., House Armed Services Committee, *Subcommittee on Tactical Air and Land Forces Hearing on Post Iraq and Afghanistan: Current and Future Roles for UAS and the Fiscal Year 2014 Budget Request,* April 23, 2013 (testimony of Dyke D. Weatherington, Director, Unmanned Warfare and Intelligence, Surveillance, and Reconnaissance, Department of Defense), https://www.gpo.gov/fdsys/pkg/CHRG-113hhrg80763/html/CHRG-113hhrg80763.htm.

were deployed to Afghanistan: John W. Lent, *480th Intelligence, Surveillance, and Reconnaissance Wing: Heritage Pamphlet* (Langley AFB, VA: 480th Intelligence, Surveillance, and Reconnaissance Wing, 2012), 121.

could watch an area more than four kilometers wide: Jackson, Lamartin, and Yahn, *WAMI Final Report,* 7.

and beam ten individual chip-outs: Sierra Nevada Corporation, *Gorgon Stare Persistent Wide Area Airborne Surveillance (WAAS) System,* pamphlet, n.d.

the aircraft flew 10,000 hours: Sierra Nevada Corporation, "Sierra Nevada Corporation Achieves Milestone for USAF's Advanced Wide-Area Airborne Persistent Surveillance System — Gorgon Stare Increment 2," press release, July 1, 2014, https://www.sncorp.com/press-releases/snc-gorgon-stare/.

Aircrews stationed in Nevada piloted the drones: NATO Standardization

Agency, *Standards Related Document ATP-3.3.7.1: UAS Tactical Pocket Guide* (Brussels, Belgium: NATO Standardization Agency, April 22, 2014), B-1.

a group of analysts downloaded the footage: Sierra Nevada Corporation, *Gorgon Stare Persistent Wide Area Airborne Surveillance (WAAS) System.*

Topps traveled to Afghanistan: Topps, interview, June 30, 2017.

53 *Big Safari took delivery of 10 complete ARGUS cameras:* Topps, interview, June 30, 2017.

The second iteration of Gorgon Stare: House Armed Services Committee, *Subcommittee on Tactical Air and Land Forces Hearing on Post Iraq and Afghanistan* (Weatherington testimony).

and a smaller wide-area infrared camera: Harris Corporation, "2014 Gorgon Stare Sensor Award," press release, May 14, 2014, https://www.harris.com/press-releases/2014/05/2014-gorgon-stare-sensor-award.

An unpublished Air Force report from 2014: Cited in "Air Force Quest for Blue Devil Replacement Centers on Gorgon Stare, ACC," Inside Defense, April 17, 2014, https://insidedefense.com/daily-news/air-force-quest-blue-devil-replacement-centers-gorgon-stare-acc.

Gorgon Stare II deployed to Afghanistan in 2014: Sierra Nevada Corporation, "Sierra Nevada Corporation Achieves Milestone."

54 *"the synoptic view":* Larry James, interview with the author, July 12, 2016.

could often track a range of suspects: Mark Cooter, interview with the author, June 1, 2017.

Sierra Nevada declined to use: Poss, interview, July 7, 2016, and Antonides, interview, February 16, 2017.

"astoundingly good": Marion, interview, February 10, 2017.

55 *In addition to seeking out insurgents and IEDs:* North Atlantic Treaty Organization, *Standards Related Document ATP-3.3.7.1: UAS Tactical Pocket Guide* (Brussels, Belgium: NATO Standardization Agency, April 2014), B-1.

alluded to having worked on: Cooter, interview, June 1, 2017.

intelligence cells working with the system: Topps, interview, June 30, 2017.

the program was redesignated: Department of the Air Force, "PRDTB3 / MQ-9 UAS Payloads," *Fiscal Year (FY) 2019 Budget Estimates,* Air Force Justification Book, Volume 2 of 2 Aircraft Procurement (Washington, DC: US Department of Defense, 2018), 1, www.saffm.hq.af.mil/Portals/84/documents/FY19/Proc/Air%20Force%20Aircraft%20Procurement%20Vol%20II%20Mods%20FY19.pdf?ver=2018-02-13-093538-670.

the Air Force deployed Gorgon Stare to Syria: Department of the Air Force, *Presentation to the House Armed Services Committee, Subcommittee on Tactical Air and Land Forces,* testimony of Lt. Gen. Arnold W. Bunch Jr. and Lt. Gen. Jerry "JD" Harris Jr., June 7, 2017, docs.house.gov/meetings/AS/AS25/20170607/106065/HHRG-115-AS25-Wstate-BunchA-20170607.pdf.

Responding to an "urgent operational need": Department of the Air Force, *Fiscal Year (FY) 2017 President's Budget Submission,* Air Force Justification

Book Volume 3b of 3, Research, Development, Test & Evaluation (Washington, DC: US Department of Defense, February 2016), www.saffm.hq.af.mil/Portals/84/documents/FY17/AFD-160208-053.pdf?ver=2016-08-24-102138-420.

an effort that was ongoing as of 2018: Rachel Cohen, "Gorgon Stare to Receive BLOS Upgrades While Air Force Explores Replacement," Inside Defense, April 6, 2018, https://insidedefense.com/daily-news/gorgon-stare-receive-blos-upgrades-while-air-force-explores-replacement.

The Air Force also worked for several years: Department of Defense, *Fiscal Year (FY) 2017 President's Budget Submission.*

can locate a large number of communications devices: US Senate, 113th Cong., 1st Sess., Report 113-44, *National Defense Authorization Act for Fiscal Year 2014, Report (to Accompany S. 1197) to Authorize Appropriations for Fiscal Year 2014 for Military Activities of the Department of Defense and for Military Construction, to Prescribe Military Personnel Strengths for Such Fiscal Year, and for Other Purposes,* June 20, 2013, https://www.gpo.gov/fdsys/pkg/CRPT-113srpt44/html/CRPT-113srpt44.htm.

described Gorgon Stare as "invaluable": US House of Representatives, 115th Cong., 1st Sess., HR Report 115-200, *Report of the Committee on Armed Services, National Defense Authorization Act for Fiscal Year 2018,* July 6, 2017, p. 35, https://www.congress.gov/115/crpt/hrpt200/CRPT-115hrpt200.pdf.

"numerous" combat units that have referred to it: Ibid.

demand for systems like Gorgon Stare will only increase in the future: Michael J. Kanaan, email to the author, March 16, 2018.

56 *"can tell where you've been and what you've been eating":* Marion, interview, June 7, 2017.

now channels all-seeing surveillance footage: Cheryl Gerber, "Video Program Expands Imagery," *Geospatial Intelligence Forum* 10, no. 7 (October 2012), https://issuu.com/kmi_media_group/docs/gif_10-7_final.

A program called Hiper Stare: "PIXIA Corp.: Meeting Interoperability Through Data Access and Dissemination," *CIO Review,* July 2014, https://geographic-information-system.cioreview.com/vendor/2014/pixia.

received an Oscar for his work: "Rahul Thakkar," Internet Movie Database. www.imdb.com/name/nm1179303/.

the NSA, which was in the midst of an ambitious effort: Nakashima and Warrick, "For NSA Chief, Terrorist Threat Drives Passion to 'Collect It All.'"

After the airplane performed impressively: Marion, Interview, June 7, 2017.

57 *Blue Devil went to Afghanistan in December 2010:* "Blue Devil," GlobalSecurity.org, June 29, 2012, https://www.globalsecurity.org/intell/systems/blue-devil.htm.

Just 280 days: Hearing on National Defense Authorization Act for Fiscal Year 2014, Before the Committee on Armed Services, Subcommittee on Intelligence, Emerging Threats and Capabilities, H.A.S.C No. 113-30, April 16, 2013 (statement of Alan R. Shaffer, Acting Assistant Secretary of Defense for Research and Engineering), 3.

Shortly thereafter, it was joined: William Ross and Mike Kelly, "Wide-Area Motion Imaging (WAMI) Technology and Systems," PowerPoint presentation, MIT Lincoln Laboratory, December 2015.

accrued over 200,000 flight hours: Logos Technologies, "Kestrel Wide-Area Motion Imagery for Aerostats," https://www.logostech.net/products-services/kestrel-wide-area-motion-imagery/.

building a surveillance aircraft equipped with ARGUS-IS: US House of Representatives, Armed Services Committee, Subcommittee on Emerging Threats and Capabilities, *Hearing on Department of Defense (DOD) Fiscal Year 2016 Science and Technology Programs: Laying the Groundwork to Maintain Technological Superiority,* March 26, 2015 (statement by Dr. Arati Prabhakar, Director, Defense Advanced Research Projects Agency [DARPA]).

58 *the Army attempted to mount ARGUS on an experimental:* Department of Defense, *Fiscal Year (FY) 2014 President's Budget Submission,* Defense Advanced Research Projects Agency, Justification Book Volume 1 of 1, Research, Development, Test & Evaluation, Defense-Wide (Washington, DC: Department of Defense, April 2013), 14.

The Army called off . . . the $297 million that it had spent developing it: W. J. Hennigan, "Army Lets Air out of Battlefield Spyship Project," *Los Angeles Times,* October 23, 2013, www.latimes.com/business/la-fi-blimp-fire-sale-20131023-story.html#axzz2ieTSRmPk.

the Air Force's attempt . . . an equally expensive disaster: Richard Whittle, "Blue Devil Airship Maker Sends SOS After Air Force Says Pack It Up," Breaking Defense, June 15, 2012, https://breakingdefense.com/2012/06/blue-devil-airship-maker-sends-sos-after-air-force-says-pack-it/.

a tiny unit so secretive . . . shared the sky with Gorgon Stare: Joseph Trevithick, "Shadowy USAF Spy Plane Spotted over Seattle Reportedly Reappears over Syria," The Drive, January 9, 2018, www.thedrive.com/the-war-zone/17511/shadowy-usaf-spy-plane-spotted-over-seattle-reportedly-reappears-over-eastern-syria.

The Air Force will say only: David Axe, "The U.S. Air Force's Most Secretive Squadron," *War Is Boring* (blog), Medium, March 16, 2014, https://medium.com/war-is-boring/the-u-s-air-forces-most-secretive-squadron-c6bacc520562.

building a fleet of surveillance planes . . . they are being used for: Scout Warrior, "High-Tech Army Spy Plane Supports Combat Ops in Africa, South America," Scout.com, May 19, 2017, www.scout.com/military/warrior/story/1762260-high-tech-army-spy-plane-supports-combat-ops.

The Army is also readying a fleet: Joseph Trevithick, "We Now Know Exactly What Sensors the Army's Powerful New RO-6A Spy Planes Will Carry," *War Zone* (blog), Drive.com, February 20, 2018, www.thedrive.com/

the-war-zone/18619/we-now-know-exactly-what-sensors-the-armys-power-
ful-new-ro-6a-spy-planes-will-carry; and "Performance-Enhanced Airborne
Reconnaissance Low (PeARL)," OGSystems, https://www.ogsystems.com/
pearl.html.

"Advanced Wide Area Motion Imagery": Office of the Secretary of Defense,
"PE 0603699D8Z / Emerging Capabilities Technology Development," *Fiscal
Year (FY) 2019 Budget Estimates,* Defense-Wide Justification Book Volume 3A
of 5, Research, Development, Test & Evaluation (Washington, DC: US Department of Defense, February 2018), 6 of 20.

59 *exploring the idea of a replacement for Gorgon Stare:* Rachel Cohen, "Gorgon Stare to Receive BLOS Upgrades While Air Force Explores Replacement,"
Inside Defense, April 6, 2018, https://insidedefense.com/daily-news/gorgon-
stare-receive-blos-upgrades-while-air-force-explores-replacement.

A laudatory webpage: "DoD Scientist of the Quarter: Mr. Daniel Uppenkamp, Air Force Research Laboratory (AFRL)," US Department of Defense,
Research and Engineering Enterprise, March 28, 2014, https://www.acq.osd.
mil/chieftechnologist/scientist/2014-4thQtr.html.

60 *the Blue Devil fleet consisted of just four aircraft:* Gabe Starosta, "USAF
Winds Down Blue Devil Program While Preparing Response to Hill," *Inside
the Air Force* 25, no. 4 (January 24, 2014): 8.

Blue Devil helped lead forces: Amy Butler, "Air Force Mulls Continued Blue
Devil 1 Ops," *Aviation Week & Space Technology,* March 18, 2013, aviationweek
.com/awin/air-force-mulls-continued-blue-devil-1-ops.

the civilian toll of that campaign has been heavy: Combined Joint Task Force,
Operation Inherent Resolve, "CJTF–OIR Monthly Civilian Casualty Report,"
press release, June 2, 2017, www.inherentresolve.mil/News/News-Releases/Ar
ticle/1200895/combined-joint-task-force-operation-inherent-resolve-monthly
-civilian-casualty/.

An inside joke among the engineers: Antoniades, interview, February 16,
2017.

4. A Murder in Baltimore

63 *had joined the thousands of violent crime cases:* Sarah Ryley et al., "Tale
of Two Cities: Even as Murders Hit Record Low in NYC, a Mountain of Cases
Languishes in Outer Boroughs as Cops Focus More Manpower on Manhattan Cases," *New York Daily News,* January 5, 2014, www.nydailynews.com/
new-york/nyc-crime/forgotten-record-murder-rate-cases-unsolved-article-
1.1566572.

64 *a single system may cost as much as $10 million:* "First Look: New High-
Tech NYPD Helicopter Takes the Fight to the Terrorists," CBS New York, June
28, 2012, newyork.cbslocal.com/2012/06/28/first-look-new-high-tech-nypd-
helicopter-takes-the-fight-to-the-terrorists/.

usually take at least 10 minutes to arrive on-scene: "In the City That Never

Sleeps," *Vertical Magazine,* April 2, 2013, https://www.verticalmag.com/fea tures/in-the-city-that-never-sleeps/.

In the aftermath of an attack: Michael Meermans, interview with the author, January 18, 2017.

65   *the Livermore team attempted to assemble an airship:* Nathan Crawford, interview with the author, February 7, 2017.

"more ubiquitous and persistent surveillance": D. M. Pennington, *Sonoma Persistent Surveillance System,* UCRL-TR-220175 (Livermore, CA: Lawrence Livermore National Laboratory, March 28, 2006), 19–20.

these efforts have accelerated: John Keller, "US Demand for ISR Technology Shifting from Military to Counter-Terrorism, Analysts Say," *Military & Aerospace Electronics,* January 21, 2014, www.militaryaerospace.com/articles/ 2014/01/frost-isr-market.html.

has pitched its peacetime version of the camera: Meermans, interview, January 18, 2017.

BAE demonstrated ARGUS for US Customs and Border Protection: Yiannis Antoniades, interview with the author, March 14, 2017.

"harbor security, large sporting events": BAE Systems, Products & Services, "Airborne Wide-Area Persistent Surveillance System (AWAPSS)," https:// www.baesystems.com/en/product/airborne-widearea-persistent-surveillance -system-awapss.

The defense firm Harris... Dunsmuir Reservoir: Dunsmuir video available on Archive.org, https://archive.org/download/DunsmuirVideo108312015/Dun smuir%20Video1%2008312015.mp4; email from Adrian Luh, Product Marketing Manager, Harris Corporation, to Jennifer Hancock, Alameda County Sheriff's Office, August 21, 2015, downloaded from Center for Human Rights and Privacy, https://www.cehrp.org/harris-aerial-surveillance-demonstrated- at-urban-shield-2015/.

66   *Harris lobbied the local sheriff:* Email from Justin McComas, Alameda County Sheriff's Office, to Adrian Luh, Harris Corporation, August 31, 2015, downloaded from Center for Human Rights and Privacy, https://www.cehrp. org/harris-aerial-surveillance-demonstrated-at-urban-shield-2015/.

maintains an airplane equipped with a 300-megapixel WAMI camera: Commuter Air Technology, *Wide Area Motion Imagery PSI Vision 3000 with MX-15,* pamphlet, n.d.

MAG Aerospace, an aerial-imaging and -surveillance services firm: Courtney Howard, "MAG and Logos Technologies Test Wide-Area Sensor for On-Demand Airborne ISR," Intelligent Aerospace, June 16, 2016, www.intel ligent-aerospace.com/articles/2016/06/mag-and-logos-technologies-test- wide-area-sensor-for-on-demand-airborne-isr.html?platform=hootsuite.

67   *"Your target cannot escape SPYDR":* SPYDR, "Your Target Cannot Escape SPYDR," https://www2.l3t.com/spydr/test/index.html

In a promotional video for CorvusEye: Exelis, "Exelis CorvusEye 1500 Wide

Area Motion Imagery," YouTube video, 5:03, October 10, 2014, https://www.
youtube.com/watch?v=Bdln0xBxo2w&t=90s.

"What is WAMI?": Logos Technologies, "What is WAMI?" YouTube video,
3:12, September 10, 2015, https://www.youtube.com/watch?v=MKX_Kzp8hfk.

Northrop Grumman's promotional video: Metro Productions, "Northrop
Grumman TS-WAMI (3064)," Vimeo video, 2016, https://vimeo.com/15242
7820.

Other companies that appear to be actively marketing and selling: Some of
these company names come from an attendee list for a proposal conference for
USDA Forest Service AFUE Solicitation #AG-024B-S-17-0005.

68 *he had developed an audacious plan for several more development cycles:*
Ross McNutt, interview with the author, July 13, 2016.

he built an eight-lens camera: Ibid.

McNutt sketched out an operating concept: Ibid.

If any crime occurred that might benefit from an aerial perspective: Ibid.

69 *His first customer . . . flights over Baltimore:* Ibid.

That fall, McNutt traveled: Ibid.

residents of the city were statistically more likely: Patrick Radden Keefe,
"The Hunt for El Chapo," *New Yorker,* May 5, 2014, https://www.newyorker.
com/magazine/2014/05/05/the-hunt-for-el-chapo.

Less than two hours . . . three separate locations along the way: McNutt, interview, July 13, 2016.

PSS also tracked a second vehicle: Ibid.

After a different murder weeks later, the team found: Ibid.

70 *McNutt and his employer had taken measures:* Ibid.

2,643 people were killed in Juárez that year: "Drug Killings Make 2010
Deadliest Year for Mexico Border City," Associated Press, January 1, 2011,
www.foxnews.com/world/2011/01/01/mexico-border-city-record-drug-kill
ings.html.

Arnold had been funding the development: Laura and John Arnold Foundation, "Public Safety Assessment: Risk Factors and Formula," 2013, www.ar
noldfoundation.org/wp-content/uploads/PSA-Risk-Factors-and-Formula.pdf.

Arnold had learned about PSS through a podcast: Tom Dart, "Eye in the
Sky: The Billionaires Funding a Surveillance Project Above Baltimore," *Guardian,* October 15, 2016, https://www.theguardian.com/world/2016/oct/15/balti
more-surveillance-john-laura-arnold-billionaires.

71 *In the lags between law enforcement jobs:* McNutt, interview, July 14, 2016.

asked McNutt to propose an American city: Kevin Rector, "Baltimore Police Declined Using Aerial Surveillance Until Big Donors Stepped Up, Emails
Show," *Baltimore Sun,* October 22, 2016, www.baltimoresun.com/news/
maryland/sun-investigates/bs-md-sun-investigates-surveillance-genesis-
20161022-story.html.

The previous year had been the bloodiest on record: Tim Prudente and Wyatt Massey, "Rapper Lor Scoota's Manager Killed Near Baltimore's Druid Hill Park," *Baltimore Sun,* July 7, 2016, www.baltimoresun.com/news/maryland/crime/bs-md-ci-fatal-shooting-20160706-story.html.

and 2015 was on track to be even worse: Kevin Rector, "Deadliest Year in Baltimore History Ends with 344 Homicides," *Baltimore Sun,* January 1, 2016, www.baltimoresun.com/news/maryland/baltimore-city/bs-md-ci-deadliest-year-20160101-story.html.

Sixty percent of the city's recent murders remained unsolved: Kevin Rector, "In 2016, Baltimore's Second-Deadliest Year on Record, Bullets Claimed Targets and Bystanders Alike," *Baltimore Sun,* January 2, 2017, www.baltimoresun.com/news/maryland/crime/bs-md-ci-homicides-2016-20170102-story.html.

The BPD had also recently completed: Nancy G. La Vigne et al., "Evaluating the Use of Public Surveillance Cameras for Crime Control and Prevention—A Summary," Urban Institute, September 2011, www.urban.org/sites/default/files/publication/27546/412401-Evaluating-the-Use-of-Public-Surveillance-Cameras-for-Crime-Control-and-Prevention-A-Summary.PDF.

it had employed StingRay: Justin Fenton, "Baltimore Police Used Secret Technology to Track Cellphones in Thousands of Cases," *Baltimore Sun,* April 9, 2015, www.baltimoresun.com/news/maryland/baltimore-city/bs-md-ci-stingray-case-20150408-story.html.

The Baltimore police had also been an early adopter of Geofeedia: Valentina Zarya, "These Hot Tech Companies Are in the CIA's Secret Investment Portfolio," *Fortune,* April 15, 2016, fortune.com/2016/04/15/cia-investment-portfolio/.

72 *Analysts ran the images through a facial-recognition system:* Geofeedia, "Baltimore County Police Department and Geofeedia Partner to Protect the Public During Freddie Gray Riots" (Case Study, Baltimore County PD), available via ACLU of Northern California, https://www.aclunc.org/docs/20161011_geofeedia_baltimore_case_study.pdf.

the major social media platforms revoked Geofeedia's access: Amina Elahi, "Geofeedia Cuts Half of Staff After Losing Access to Twitter, Facebook," *Chicago Tribune,* November 21, 2016, www.chicagotribune.com/bluesky/originals/ct-geofeedia-cuts-jobs-surveillance-bsi-20161121-story.html.

BPD had agreed to deploy software newly developed by: Sarah Gantz, "Baltimore's ZeroFox Faces Backlash over Riot Threat Report; CEO James Foster Responds," *Baltimore Business Journal,* August 4, 2015, https://www.bizjournals.com/baltimore/blog/cyberbizblog/2015/08/baltimores-zerofox-faces-backlash-over-riot-threat.html?ana%3Dtwt.

McNutt had been courting the director: Brandon Soderberg, "Persistent Transparency: Baltimore Surveillance Plane Documents Reveal Ignored Pleas to Go Public, Who Knew About the Program, and Differing Opinions on Privacy," *City Paper,* November 1, 2016, http://www.citypaper.com/news/mobtownbeat/bcp-110216-mobs-aerial-surveillance-20161101-story.html.

John Arnold made a personal donation of $360,000: Rector, "Baltimore Police Declined Using Aerial Surveillance Until Big Donors Stepped Up."

Under the terms of its agreement with the police: McNutt, Interview, August 23, 2016.

73 *on any given day. The computer would generate:* Police Foundation, *A Review of the Baltimore Police Department's Use of Persistent Surveillance (Baltimore Community Support Program)* (Washington, DC: Police Foundation, January 30, 2017), 10.

McNutt was instructed to focus: McNutt, interview, July 13, 2016.

"I'm dealing with a murder, a mugging": McNutt, interview, June 27, 2016.

74 *Rather than hiring ex-military intelligence analysts:* McNutt, interview, July 13, 2016.

The analysts found that vehicles: Ibid.

A detective involved in one of the cases: Police Foundation, *A Review of the Baltimore Police Department's Use of Persistent Surveillance,* 16.

In another instance... arrested that same day: McNutt, interview, July 13, 2016.

75 *"Bank robbers put a lot of effort into planning their escape":* Steven Suddarth, unpublished manuscript, "Suddarth WAMI Book outline."

investigators would use information about a suspect's whereabouts: McNutt, interview, July 13, 2016.

McNutt's team tracked two dozen vehicles: McNutt, interview, July 14, 2016.

After several months of operations in Baltimore: McNutt, interview, July 13, 2016.

the team was staking out one location in particular: Ibid.

76 *the Baltimore Police Department's CCTV data-retention policy:* Jay Stanley, "Baltimore Aerial Surveillance Program Retained Data Despite 45-Day Privacy Policy Limit," *Free Future* (blog), ACLU, October 25, 2016, https://www.aclu.org/blog/free-future/baltimore-aerial-surveillance-program-retained-data-despite-45-day-privacy-policy.

because the camera will invariably capture evidence: McNutt, interview, July 14, 2016.

at least seven such special operations: Police Foundation, *A Review of the Baltimore Police Department's Use of Persistent Surveillance,* 15.

McNutt believed that the convenience store was selling ingredients: McNutt, interview, July 13, 2016.

had approved the aerial stakeout: McNutt, interview, July 14, 2016.

77 *spring of 1998... "we tried":* [Redacted], email to [Redacted], April 23, 1998, and [Redacted], email to [Redacted], April 23, 2016. These emails were released as part of a response to a Freedom of Information Act request from BuzzFeed Politics on January 7, 2016.

78 *PV Labs, a Canadian firm, has flown:* John Bastedo, interview with the

author, June 14, 2017, and Christian Stork and Peter Aldhous, "This Shadowy Company Is Flying Spy Planes over US Cities," BuzzFeed News, August 4, 2017, https://www.buzzfeed.com/christianstork/spy-planes-over-american-cities.

The Australian Department of Defence has tested a wide-area camera: Nigel Pittaway, "Five Eyes Test New Tech in Exercise for Reducing Urban Combat Risks," C4ISRNET, November 30, 2017, https://www.c4isrnet.com/intel-geo int/sensors/2017/11/30/five-eyes-test-new-tech-in-exercise-for-reducing-ur ban-combat-risks/; and David Pugliese, "Defence Scientists in Montreal to Test Technologies for Fighting in Urban Areas," *Ottawa Citizen,* September 2, 2018, https://ottawacitizen.com/news/national/defence-watch/defence-scien tists-in-montreal-to-test-technologies-for-fighting-in-urban-areas.

has flown its giant surveillance systems over downtown Boston: William Ross and Mike Kelly, "Wide-Area Motion Imaging (WAMI) Technology and Systems," PowerPoint presentation, MIT Lincoln Laboratory, December 2015.

have been used by researchers at the University of Missouri: 46th Annual IEEE AIPR 2017, Big Data, Analytics, and Beyond, Washington, DC, October 10–12, 2017, 2017 *Applied Imagery Pattern Recognition Workshop* (proceedings), p. 39, www.aipr-workshop.org/images/2017/AIPR_Program_Booklet_2017.pdf.

In Pasadena, he flew a test flight: Suddarth, email, August 8, 2018.

part of a $4 million exercise backed by the CIA and NSA: US Department of Defense, OSD RDT&E Budget Item Justification (R2a Exhibit), "0603826D8Z —Quick Reactions Special Projects (QRSP)," February 2008, www.dtic.mil/ descriptivesum/Y2009/OSD/0603826D8Z.pdf; and Glenn Fogg, "How to Bet ter Support the Need for Quick Reaction Capabilities in an Irregular Warfare Environment," PowerPoint presentation, US Department of Defense, Rapid Reaction Technology Office, Washington, DC, April 21, 2009.

the people of Lubbock, Texas, were surveilled by: GlobalSecurity.org, "Wide Area Persistent Surveillance (WAPS)," n.d., https://www.globalsecurity.org/ intell/systems/waps.htm; and unpublished notes provided to the author.

describes it as a "large" data set: Andrew Ladas, "ISR Concepts for Unique Persistent Wide Area Motion Imagery Dataset," PowerPoint presentation, US Army RDECOM, Aberdeen, MD.

The CIA proposed conducting a follow-up to Bluegrass: Office of the Sec retary of Defense, *Quick Reactions Special Projects (QRSP),* PE 0603826D8Z (Washington, DC: US Department of Defense, February 2014), 15.

79 *In the summer of 2017:* Tyler Rogoway and Joseph Trevithick, "This Mys terious Military Spy Plane Has Been Flying Circles over Seattle for Days," *War Zone* (blog), The Drive, August 3, 2017, www.thedrive.com/the-war zone/13154/this-mysterious-military-spy-plane-has-been-flying-circles-over seattle-for-days.

flew a mock surveillance mission: Email from Nathan Crawford, February 26, 2018.

In two cases from the 1980s: California v. Ciraolo, 476 U.S. 207 (1986), and Florida v. Riley, 488 U.S. 445 (1989).

In another case, Kyollo v. United States: Kyllo v. United States, 533 U.S. 27 (2001).

80 *"It's pretty much a slam-dunk case":* McNutt, interview, July 13, 2016.

Since the PSS cameras were not peering: Ibid.

On July 23, 2014, a Reddit user . . . surveillance operations: "Plane Circling over McLean/Langley Area Last Few Days," Reddit, https://www.reddit.com/r/nova/comments/2bgj1p/plane_circling_over_mcleanlangley_area_last_few/.

81 *Another aircraft, tail number N859JA:* Lawrence Harmon, "FBI Too Quiet on Quincy Planes," *Boston Globe,* May 17, 2013, http://www.bostonglobe.com/opinion/2013/05/16/fbi-too-quiet-quincy-planes/0h9EObhcoQvh41Wxdp AHZM/story.html.

"Anyone know who has been flying the light plane": @scanbaltimore on Twitter, May 2, 2015, https://twitter.com/scanbaltimore/status/59467121402883 6864.

soon discovered that a second aircraft: Craig Timberg, "Surveillance Planes Spotted in the Sky for Days After West Baltimore Rioting," *Washington Post,* May 5, 2015, https://www.washingtonpost.com/business/technology/surveillance-planes-spotted-in-the-sky-for-days-after-west-baltimore-rioting/2015/05/05/c57c53b6-f352-11e4-84a6-6d7c67c50db0_story.html?utm_term=.389d2 dec6c2a.

filed a Freedom of Information Act request: Jay Stanley, "Mysterious Planes over Baltimore Spark Surveillance Suspicions," MSNBC.com, May 6, 2015, www.msnbc.com/msnbc/mysterious-planes-over-baltimore-spark-surveil lance-suspicions.

Evidence logs released in response: Memorandum, "(U) Baltimore Police Department, Civil Unrest — Riots, Domestic Police Cooperation April 27, 2015," Federal Bureau of Investigation, May 1, 2015, documents released in response to a Freedom of Information Act request submitted by the American Civil Liberties Union, available at https://archive.org/stream/FBI-Baltimore-Freddie-Gray/fbi_memo_and_evidence_logs_djvu.txt.

It had used both daylight and infrared soda-straw cameras: Nathan Freed Wessler, "FBI Documents Reveal New Information on Baltimore Surveillance Flights," ACLU, October 30, 2015, https://www.aclu.org/blog/privacy-technology/surveillance-technologies/fbi-documents-reveal-new-informa tion-baltimore.

NG Research turned out to be an FBI front company: Ibid.

the FBI also provided an excerpt: Federal Bureau of Investigation, *FBI Domestic Investigations and Operations Guide,* available at https://vault.fbi.gov/FBI%20Domestic%20Investigations%20and%20Operations%20Guide%20%28DIOG%29?.

A former FBI agent who ran a special-operations group: James Schweitzer, interview with the author, May 19, 2017.

In testimony before the House Committee on the Judiciary: Oversight of the Federal Bureau of Investigation: Hearing Before the Committee of the Judiciary, House of Representatives, 114th Cong., October 22, 2015 (testimony of James Comey, Director, Federal Bureau of Investigation), https://www.justice.gov/sites/default/files/testimonies/witnesses/attachments/2016/01/29/10-22-15_fbi_comey_testimony_re_oversight_of_the_federal_bureau_of_investigation_web_ready.pdf.

82 *BuzzFeed reported that the FBI had used more than 100 aircraft:* Peter Aldhous and Charles Seife, "Spies in the Skies," BuzzFeed News, April 6, 2016, https://www.buzzfeed.com/peteraldhous/spies-in-the-skies.

Most of these flights had been operated: Jack Gillum, Eileen Sullivan, and Eric Tucker, "FBI Behind Mysterious Surveillance Aircraft over US Cities," Associated Press, June 2, 2015, https://www.msn.com/en-us/news/us/fbi-behind-mysterious-surveillance-aircraft-over-us-cities.

The FBI explained . . . stopped almost entirely: Aldhous and Seife, "Spies in the Skies."

The Department of Homeland Security Operates: US Department of Homeland Security, Privacy Impact Assessment Update for the Aircraft Systems, DHS/CBP/PIA-018(a), April 6, 2018, 5, https://www.dhs.gov/sites/default/files/publications/privacy-pia-cbp018a-aircraftsystems-april2018.pdf.

83 *the agency hosted a demonstration:* Logos Technologies, "Logos Technologies Demonstrates Kestrel for U.S. Border Security," press release, March 29, 2012, https://www.logostech.net/logos-technologies-demonstrates-kestrel-for-u-s-border-security/.

the flow of people moving through the area: Marion, interview, June 7, 2017.

Two of CBP's Reapers are already equipped . . . pounds of cocaine: US Department of Homeland Security, "Written Testimony of CBP Office of Air and Marine Assistant Commissioner Randolph Alles, CBP Office of Technology Innovation and Acquisition Assistant Commissioner Borkowski, and CBP Office of Border Patrol Deputy Chief Ron Vitiello for a Senate Committee on Homeland Security and Governmental Affairs for a Hearing Titled 'Securing the Border: Fencing, Infrastructure, and Technology Force Multipliers,'" release date May 13, 2015, https://www.dhs.gov/news/2015/05/13/written-testimony-cbp-senate-committee-homeland-security-and-governmental-affairs.

CBP has already invested: Vijayan Asari, interview with the author, March 14, 2017, and Lloyd L. Coulter et al., "Repeat Station Imaging for Rapid Airborne Change Detection," in *Time-Sensitive Remote Sensing,* edited by Christopher D. Lippitt, Douglas A Stow, and Lloyd L. Coulter (New York: Springer, 2015), 39.

one Department of Homeland Security document from 2013: US Department of Homeland Security, Privacy Impact Assessment for the Aircraft Systems,

DHS/CBP/PIA-018, September 9, 2013, https://www.dhs.gov/sites/default/
files/publications/privacy-pia-cbp-aircraft-systems-20130926.pdf.

the Marshals Service maintains a secretive task force: US Marshals Service,
"Technical Operations Group," https://www.usmarshals.gov/investigations/
tog/tog.htm.

In 2017 reporters from BuzzFeed: Peter Aldhous, "We Trained a Computer
to Search for Hidden Spy Planes. This Is What It Found," BuzzFeed News, August 7, 2017, https://www.buzzfeed.com/peteraldhous/hidden-spy-planes.

84   *An investigation published by the* Texas Observer: G. W. Schulz and Me-
lissa del Bosque, "The Eyes Above Texas," *Texas Observer,* May 23, 2018,
https://www.texasobserver.org/planes/.

None of these operations have been widely publicized: Aldhous, "We
Trained a Computer to Search for Hidden Spy Planes."

there were more than 600 state and local law enforcement agencies: Dan Gettinger, "Public Safety Drones: An Update," Center for the Study of the Drone,
May 28, 2018, https://dronecenter.bard.edu/public-safety-drones-update/.

In the summer of 2011 a Department of Homeland Security Reaper: "Predator Drone Helps Convict North Dakota Farmer in First Case of Its kind," Fox
News, January 28, 2014, www.foxnews.com/us/2014/01/28/first-american-
gets-prison-with-assistance-predator-drone.html.

The California National Guard has, in turn, deployed Predators: California National Guard, "Cal Guard Launches Remotely Piloted Aircraft to Search
El Dorado National Forest for Motorcyclist," press release, July 30, 2015,
http://www.nationalguard.mil/News/Article-View/Article/611476/cal-guard-
launches-remotely-piloted-aircraft-to-search-el-dorado-national-fores/;   and
Army Staff Sgt. Edward Siguenza, "California Air Guard Drone Helps Authorities Fight Carr Fire," US Department of Defense, California National Guard,
July 31, 2018, https://dod.defense.gov/News/Article/Article/1588830/california
-air-guard-drone-helps-authorities-fight-carr-fire/.

85   *as one Pentagon planning document put it:* UAS Task Force, Airspace Integration Integrated Product Team, *Unmanned Aircraft System Airspace Integration
Plan*, Version 2.0 (Washington, DC: Department of Defense, March 2011), 1.

company officials say: Patrick Tucker, "Look for Military Drones to Begin
Replacing Police Helicopters by 2025," Defense One, August 28, 2017, www.
defenseone.com/technology/2017/08/look-military-drones-replace-police-he
licopters-2025/140588/.

In 2018 a variant of the Reaper: David Szondy, "SkyGuardian Drone Completes Historic Transatlantic Flight," New Atlas, July 11, 2018, https://newatlas.
com/skyguardian-transatlantic-landing/55420/.

87   *other BPD investigators had declined to act on evidence:* Police Foundation, *A Review of the Baltimore Police Department's Use of Persistent Surveillance*, 14.

88   *BPD would later claim ... beyond the police commissioner:* Kevin Rector,

"Cummings: Commissioner Davis 'Apologized Profusely' for Not Disclosing Surveillance Program," *Baltimore Sun,* September 2, 2016, www.baltimoresun.com/news/maryland/crime/bs-md-ci-cummings-davis-meeting-2016 0902-story.html.

Even the mayor was kept in the dark: Kevin Rector and Luke Broadwater, "Report of Secret Aerial Surveillance by Baltimore Police Prompts Questions, Outrage," *Baltimore Sun,* August 24, 2016, www.baltimoresun.com/news/maryland/baltimore-city/bs-md-ci-secret-surveillance-20160824-story.html.

In 2016 it was revealed: Dustin Slaughter, "This Isn't a Google Streetview Car, It's a Government Spy Truck," Motherboard, May 12, 2016, https://motherboard.vice.com/en_us/article/bmvjwm/this-isnt-a-google-streetview-car-its-a-government-spy-truck.

a technique known as "parallel construction": Human Rights Watch, *Dark Side: Secret Origins of Evidence in US Criminal Cases* (New York: Human Rights Watch, 2017), 5.

police departments have even opted to drop charges: Cyrus Farivar, "FBI Would Rather Prosecutors Drop Cases Than Disclose Stingray Details," Ars Technica, April 7, 2015, https://arstechnica.com/tech-policy/2015/04/fbi-would-rather-prosecutors-drop-cases-than-disclose-stingray-details/.

89 *When the story broke, the police sergeant:* G. W. Schulz and Amanda Pike, "Hollywood-Style Surveillance Technology Inches Closer to Reality," Center for Investigative Reporting, April 11, 2014, available via Internet Archive, https://web.archive.org/web/20140416141623/cironline.org/reports/hollywood-style-surveillance-technology-inches-closer-reality-6228.

In 2018 Consolidated Resource Imaging was awarded: Nathan Crawford, interview with the author, March 1, 2018.

McNutt told me...could be easily convinced: McNutt, interview, July 14, 2016.

5. Pixels into Ploughshares

90 *"has to serve a positive purpose":* Nathan Crawford, interview with the author, January 26, 2017.

In the summer of 2008...acceptance speech: Ross McNutt, "Wide Area Surveillance and Counter Narco-Terrorism Operations," PowerPoint presentation, March 6, 2012, downloaded from https://www.slideshare.net/Shadowairs/waass-pss-counter-narco-terrorism-briefing.

In 2010 Consolidated Resource Imaging: Frank Colucci, "Persistence on Patrol," *Avionics,* May 1, 2013.

91 *Over the course of the response to Hurricane Katrina:* United States Coast Guard, "The U.S. Coast Guard & Hurricane Katrina," https://www.uscg.mil/history/katrina/katrinaindex.asp.

In one exercise off the Scottish coast in 2016: Insitu, "Insitu ScanEagle Completes Successful Maritime Surface Search at Royal Navy's Unmanned Warrior," press release, November 15, 2016, https://www.insitu.com/press-re

leases/Insitu-ScanEagle-Completes-Successful-Maritime-Surface-Search-at-Royal-Navys-Unmanned-Warrior.

In 2013 Sierra Nevada Corporation flew: Meermans, interview, January 18, 2017.

Persistent Surveillance Systems has carried out similar operations: McNutt, "Wide Area Surveillance in Support of Law Enforcement," PowerPoint presentation, January 2014, downloaded from https://info.publicintelligence.net/PSS-WideAreaSurveillance.pdf.

In 2018 the Indiana National Guard announced: Lonnie Wiram, "Mission Impassable," Indiana Air National Guard, 181st Intelligence Wing Public Affairs, September 19, 2018, https://www.dvidshub.net/news/295390/mission-impassible.

The Australian Department of Defence optimized: Australian Government, Department of Defence, "WASABI — Angel Fire," https://www.dst.defence.gov.au/projects/wasabi-angel-fire.

92 *In his world, the aircraft that drop fire suppressant . . . "game changer":* Zachary Holder, interview with the author, April 6, 2017.

the Forest Service awarded CRI: "FedBizOpps Issue of December 14, 2017, FBO #5865 Award," FBO Daily, December 12, 2017, www.fbodaily.com/archive/2017/12-December/14-Dec-2017/FBO-04764027.htm.

lease a new infrared camera 10 times larger: Crawford, interview, March 1, 2018.

Holder estimates that the service would need only five airplanes: Holder, interview, April 6, 2017.

93 *While conducting overwatch for the Coca-Cola 600 NASCAR race . . . $250,000 in profit:* Ross McNutt, "Wide Area Surveillance in Support of Law Enforcement," PowerPoint presentation, January 2014, and McNutt, interview, July 13, 2016.

94 *two researchers at Virginia Tech:* Kathleen L. Hancock and Md Rauful Islam, *Final Report: Use of Wide-Area Motion Imagery (WAMI) for Transportation Planning and Operations,* VT 2013-03 (Blacksburg, VA: Virginia Tech Transportation Institute, February 2016), 4–5.

Over the course of the surveillance program in Baltimore: Police Foundation, *A Review of the Baltimore Police Department's Use of Persistent Surveillance,* 15.

In 10 cases, analysts were able to identify: Ibid.

95 *in each case the camera's cost:* Meermans, interview, January 18, 2017.

Gorgon Stare, which costs about $20 million: US Air Force, Exhibit P-40, Budget Line Item Justification: PB 2019, "PRDTB3 / MQ-9 UAS Payloads," February 2018, p. 1, www.dtic.mil/procurement/Y2019/AirForce/stamped/U_P40_PRDTB3_BSA-5_BA-5_APP-3010F_PB_2019.pdf.

could be owned and operated by state and federal organizations: Brian Leininger, interview with the author, April 7, 2017.

96 *In one such operation, in 2016:* Stork and Aldhous, "This Shadowy Company Is Flying Spy Planes over US Cities."
the insurance industry has also explored the idea: Crawford, interview, January 26, 2017.
the citizens of Wilmington, North Carolina, became: Stork and Aldhous, "This Shadowy Company Is Flying Spy Planes over US Cities."
The company believes that it could sell this information to retailers: Ibid.
A laboratory at the University of Dayton is developing: Vijayan Asari, Director, Vision Lab, University of Dayton, interview with the author, March 14, 2017.

6. What's Next for Wide-Area Surveillance

98 *"No. When it comes to the world of actually collecting information":* Michael Meermans, interview with the author, August 5, 2016.

99 *a separate DARPA program called AWARE:* Patrick Llull et al., "Characterization of the AWARE 40 Wide-Field-of-View Visible Imager," *Optica* 2, no. 12 (December 2015): 1086–89.
The individual pixels on the cell phone chips: Yiannis Antoniades, interview with the author, March 14, 2017; and Richard Sinn, "Virtual Pan-Tilt-Zoom for a Wide-Area-Video Surveillance System" (master's thesis, Department of Electrical Engineering and Computer Science, MIT, August 22, 2008), 21.
which is the size of the pixels in an iPhone 8 camera: Daniel Yang, Stacy Wegner, and Ray Fontaine, "Apple iPhone 8 Plus Teardown," Tech Insights, September 10, 2017, www.techinsights.com/about-techinsights/overview/blog/apple-iphone-8-teardown/.
you could, theoretically, build a Gorgon Stare: Antoniades, interview, March 14, 2017.
In 2012 it took one of Livermore's: Sheila Vaidya, interview with the author, February 7, 2017.
The first wide-area-surveillance cameras alone: Dwayne Jackson, David Lamartin, and Jacqueline Yahn, *WAMI Final Report* (Washington, DC: Department of Defense, Secretary of the Air Force for Acquisition, December 11, 2012), 21.

100 *The Marine Corps is preparing:* US Navy, "PE 0305242M / (U)Unmanned Aerial Systems (UAS) Payloads," February 2018, 3 of 29; and "Tactical Nighttime Wide Area Surveillance," GovTribe, last updated October 7, 2011, https://govtribe.com/project/tactical-nighttime-wide-area-surveillance-1/activity.
Logos Technologies' Redkite: Richard Tomkins, "Wide-Area Sensor Flight-Tested on Small Drone," UPI, February 15, 2017, www.upi.com/Defense-News/2017/02/15/Wide-area-sensor-flight-tested-on-small-drone/1541487169770/.
Eventually, though, attempts to further miniaturize pixels: Antoniades, interview, March 14, 2017.

Agile Spotter, a 3.2-megapixel unit: John Marion, interview with the author, May 2, 2018.

A real-life superresolution tool: University of Dayton School of Engineering Vision Lab, "Visibility Improvement," available through Internet Archive at https://udayton.edu/engineering/centers/vision_lab/research/wide_area_sur veillance/visibility_improvements.php.

101 *targets become easier to track:* Asari, interview, March 14, 2017.

something called the digital-pixel focal plane array: William Ross, interview with the author, February 2, 2018, and email to the author, May 8, 2018.

developed by the MIT Lincoln Laboratory: Dan Cress, interview with the author, October 26, 2016.

enabling it to monitor an area the size of Pittsburgh: Ross and Kelly, "Wide-Area Motion Imaging (WAMI) Technology and Systems," and "Wide-Area Infrared Sensor for Persistent Surveillance (WISP)," MIT Lincoln Laboratory, available through Internet Archive at https://www.ll.mit.edu/mission/elec tronics/ait/digital-pixel-fpa/wide-area-infrared-sensor.html.

each pixel can also correct motion . . . an ordinary pixel: Kenneth I. Schultz et al., "Digital-Pixel Focal Plane Array Technology," *Lincoln Laboratory Journal* 20, no. 2 (2014): 36–51.

Redkite was named "Best New Product": Logos Technologies, "Logos Technologies Wins Coveted Aviation Week Award for Redkite Wide-Area Sensor," press release, December 5, 2017, https://www.logostech.net/logos-wins-avia tion-week-award-redkite-sensor/.

103 *As of the spring of 2018, Logos:* Marion, interview, May 2, 2018.

Harris has been pitching the CorvusEye 1500: "Wide Area Motion Imagery," *ESD Spotlight* no. 13 (2015): 5; and "Seeing the Big Picture," *Eurosatory Daily,* June 17, 2016.

it could nevertheless be purchased by up to 47 foreign governments: Gordon Arthur, "EO/IR Special Report: CorvusEye Scans the Region," Shephard Media, February 12, 2018, https://www.shephardmedia.com/news/digital-bat tlespace/eoir-special-report-corvuseye-scans-region/.

a 300-megapixel camera that the company notes: PV Labs, "PV Labs Intelligent Imaging," www.pv-labs.com.

At least one European airborne-surveillance firm: Vigilance, "Wide Area Persistent Surveillance," www.vigilance.nl/wide-area-persistent-surveillance. html.

The Israeli defense firm Elbit Systems . . . such events: Barbara Opall-Rome, "Elbit Unveils Time-Traveling Airborne Surveillance System," *Defense News,* June 12, 2017, www.defensenews.com/articles/elbit-unveils-time-traveling-airborne-surveillance-system.

The UK Ministry of Defence has quietly: Mike Davies, Yvan Petillot, and Janet Forbes, *Mid Term Review — Signal Processing for a Networked Battlespace* (Edinburgh, Scotland: Edinburgh Consortium, Joint Research Institute of Signal Image Processing, University Defence Research Collaboration, September 30, 2015), 21.

test flights tracking cars along major roadways: R. I. Young and S. B. Foulkes, "Towards a Real-Time Wide Area Motion Imagery System," *SPIE Proceedings* 9652 (2015), Optics and Photonics for Counterterrorism, Crime Fighting, and Defence 11, and Optical Materials and Biomaterials in Security and Defense Systems Technology 12, doi: 10.1117/12.2194568, pp. 8–9.

104 *The German Aerospace Center is developing:* DLR Center for Satellite Based Crisis Information, "ARGOS — Airborne Wide Area High Altitude Monitoring System," available through Internet Archive at https://www.zki.dlr.de/project/106.

governments of Germany: See, for example, Mickael Cormier, Lars Wilko Sommer, and Michael Teutsch, "Low Resolution Vehicle Re-identification Based on Appearance Features for Wide Area Motion Imagery," *2016 IEEE Winter Applications of Computer Vision Workshops (WACVW),* 1–7, doi: 10.1109/WACVW.2016.7470114.

and the Netherlands: Jasper van Huis et al., "Vehicle-Tracking in Wide-Area Motion Imagery from an Airborne Platform," SPIE Newsroom, December 8, 2015, spie.org/newsroom/6212-vehicle-tracking-in-wide-area-motion-imagery-from-an-airborne-platform, doi: 10.1117/2.1201511.006212.

Singapore is working: IPI Singapore, "The Future of City Monitoring: Automated Surveillance from the Sky," https://www.ipi-singapore.org/technology-offers/future-city-monitoring-automated-surveillance-sky.

Farther south, the Australian Department of Defence: Australian Government, Department of Defence, "Defence Experimentation Airborne Platform (DEAP)," https://www.dst.defence.gov.au/research-facility/defence-experimentation-airborne-platform-deap.

engineers are using the system to develop: Australian Government, Department of Defence, "WASABI — ANGEL FIRE."

In 2017 and 2018 it brought the aircraft to Contested Urban Environment: Nigel Pittaway, "Five Eyes Test New Tech in Exercise for Reducing Urban Combat Risks," C4ISRNET, November 30, 2017, https://www.c4isrnet.com/intel-geoint/sensors/2017/11/30/five-eyes-test-new-tech-in-exercise-for-reducing-urban-combat-risks/, and David Pugliese, "Defence Scientists in Montreal to Test Technologies for Fighting in Urban Areas," *Ottawa Citizen,* September 2, 2018, https://ottawacitizen.com/news/national/defence-watch/defence-scientists-in-montreal-to-test-technologies-for-fighting-in-urban-areas.

The Pentagon also appears to be sharing its work: Jack Shanahan, "Algorithmic Warfare Cross-Functional Team (AWCFT) aka Project Maven," PowerPoint presentation, US Department of Defense, Washington, DC, October 26, 2017.

General Atomics Aeronautical Systems, the American drone maker: General Atomics Aeronautical Systems, Inc., Canberra, Australia, "GA-ASI Expands Team Reaper Australia," press release, September 3, 2018, https://www.businesswire.com/news/home/20180903005033/en/GA-ASI-Expands-Team-Reaper-Australia.

In 2015 the Russian hacking group Fancy Bear . . . Project Maven: Jeff Donn, Desmond Butler, and Raphael Satter, "Drones to Cloud Computing: AP Exposes Russian Wish List," Associated Press, February 7, 2018, www.bdtonline.com/news/drones-to-cloud-computing-ap-exposes-russian-wish-list/article_235754e4-bff5-5c57-b144-f7ebcc5549c8.html, and Kaveh Waddell, "AI Company Working on DOD Project Reportedly Breached," Axios, June 13, 2018, https://www.axios.com/ai-company-working-on-project-maven-reportedly-breached-1528930694-5ac58e86-e198-4e9a-84fd-971c467a05b4.html.

105 *China's surveillance-technology sector:* "Global and China Sensor Industry Report, 2015–2018," ReportBuyer, September 29, 2015, www.prnewswire.com/news-releases/global-and-china-sensor-industry-report-2015-2018-300151132.html.

Hangzhou Hikvision Digital Technology: Xiao Yu, "Is the World's Biggest Surveillance Camera Maker Sending Footage to China?" Voice of America, November 21, 2016, www.voanews.com/a/hikvision-surveillance-cameras-us-embassy-kabuk/3605715.html.

Scientists from the National Laboratory of Pattern Recognition at the Institute of Automation: See, for example, Xinchu Shi et al., "Using Maximum Consistency Context for Multiple Target Association in Wide Area Traffic Scenes," *2013 IEEE International Conference on Acoustics, Speech and Signal Processing,* doi: 10.1109/ICASSP.2013.6638042.

with 200 million CCTV: Paul Mozur in "Playing Cat and Mouse with a Surveillance State," *New York Times,* February 28, 2019.

China is fast becoming the world's largest: Adam Rawnsley, "Meet China's Killer Drones," *Foreign Policy,* January 14, 2016, foreignpolicy.com/2016/01/14/meet-chinas-killer-drones/.

Jordan only turned to China: Judson Berger, "Obama Administration Denied Predator Drone Request for Jordan, Rep Urges Reversal," Fox News, February 6, 2015, www.foxnews.com/politics/2015/02/06/obama-administration-denied-request-for-jordan-to-get-predator-drones-rep-urges.html.

106 *ominous portent of this future is the DragonFly:* Ross and Kelly, "Wide-Area Motion Imaging (WAMI) Technology and Systems."

The DragonFly was tested alongside: Pittaway, "Five Eyes Test New Tech in Exercise for Reducing Urban Combat Risks."

has built a drone-mounted 30-megapixel . . . even when it's completely overcast: Suddarth, interview, January 9, 2016, and materials provided to the author.

under a program called Perdix: Gideon Grudo, "Perdix Program Could Be DOD's Pathfinder to Progressive Projects," *Air Force Magazine,* June 23, 2017, www.airforcemag.com/Features/Pages/2017/June%202017/Perdix-Program-Could-be-DODs-Pathfinder-to-Progressive-Projects.aspx.

107 *I was told—firmly, but politely:* Michael J. Kanaan, email to the author, March 16, 2018.

7. It Takes a Million Eyes

108 *it took thousands of photographic interpreters:* [Redacted], National Photographic Interpretation Center, *Volume 1: Antecedents and Early Years, 1952–56,* CIA Directorate of Science and Technology, 1972, 190.

A single soda-straw-view Reaper requires eight: US Air Force, "Air Force Distributed Common Ground System," October 13, 2015, www.af.mil/About-Us/Fact-Sheets/Display/Article/104525/air-force-distributed-common-ground-system/.

According to Steven K. Rogers, a senior Air Force scientist, they are told that they must ask permission: Steven K. Rogers, speaking at GEOINT Symposium 2015, quoted in Patrick Tucker, "Robots Won't Be Taking These Military Jobs Anytime Soon," Defense One, June 22, 2015, https://www.defenseone.com/technology/2015/06/robots-wont-be-taking-these-military-jobs-anytime-soon/116017/.

filled so many high-capacity Hitachi DeskStar: Crawford, interview, January 26, 2017.

109 *the data repository generated by Constant Hawk was one of the densest:* James A. Ratches, Richard Chait, and John W. Lyons, *Some Recent Sensor-Related Army Critical Technology Events* (Washington, DC: Center for Technology and National Security Policy, National Defense University, February 2013), 12.

"It really takes a million people to watch a million people": Crawford, interview, January 26, 2017.

If you wanted to display every single pixel from a Gorgon Stare: Brian Leininger, "Autonomous Real-Time Ubiquitous Surveillance—Imaging System (ARGUS-IS)," PowerPoint presentation, The Villages Science and Technology Club Meeting, The Villages, Florida, February 8, 2016.

A highly caffeinated team of 20 analysts: Matthew Beinart, "Drones to Get New AI Algorithms Under Project Maven," *Avionics,* November 2, 2017, www.aviationtoday.com/2017/11/02/drones-get-new-ai-algorithms-project-maven/.

110 *large volumes of significant information will wind up on the cutting room floor:* Richard Nichols, interview with the author, February 24, 2017.

it even put the Jasons on the case: Jason, *Data Analysis Challenges* (McClean, VA: JASON, MITRE Corporation, December 2008), 86.

calculated that the service would need 117,000 analysts: Lance Menthe et al., *The Future of Air Force Motion Imagery Exploitation: Lessons from the Commercial World* (Santa Monica, CA: RAND Corporation, 2012), 8.

"the most boring video game in the history of the world": Alex Pasternack, "Can 'Good Kill' Make the Public Care About Drone Warfare?," Motherboard, May 19, 2015, https://motherboard.vice.com/en_us/article/gvyeg4/can-good-kill-make-the-public-care-about-the-drone-war.

111 *the Joint Forces Command paid Lockheed Martin:* Paul Richfield, "Intell Video Moves to a Netflix Model," GCN, April 6, 2011, https://gcn.com/articles/2011/03/29/c4isr-1-battlefield-full-motion-video.aspx.

it requires the footage to be meticulously tagged by hand: James Poss, interview with the author, August 24, 2016.

RAND researchers began to study reality TV: Menthe et al., *The Future of Air Force Motion Imagery Exploitation*, 15.

acknowledges two TV producers: Katie Drummond, *"Jersey Shore,* Meet Robot War: How Reality TV Could Fix Drone Data Glut," *Wired,* June 5, 2012, www.wired.com/2012/06/reality-tv-drone/.

Google Street View's algorithms: "How Google Cracked House Number Identification in Street View," *MIT Technology Review,* January 6, 2014, https://www.technologyreview.com/s/523326/how-google-cracked-house-number-identification-in-street-view/.

112 *Self-driving cars can:* Danny Shapiro, "Here's How Deep Learning Will Accelerate Self-Driving Cars," NVIDIA blog, February 24, 2015, https://blogs.nvidia.com/blog/2015/02/24/deep-learning-drive/.

In 2015 two researchers from Michigan State and Pennsylvania State: Raechel A. Bianchetti and Alan M. MacEachren, "Cognitive Themes Emerging from Air Photo Interpretation Texts Published to 1960," *International Journal of Geo-Information* 4, no. 2 (2015): 565.

113 *Many of the Gorgon Stare analysts:* Mark Cooter, interview with the author, June 1, 2017.

114 *the analysis of footage from a mission:* Patrick Biltgen and Stephen Ryan, *Activity-Based Intelligence: Principles and Applications* (Norwood, MA: Artech House, 2015), 210.

Bill Ross, the Lincoln Laboratory engineer, spent years: William Ross, interview with the author, December 14, 2016.

Suddarth had worked in artificial intelligence and machine vision: Steven Suddarth, interview with the author, September 20, 2016.

a sugar-cube-sized computer that identified incoming cruise missiles: Suraphol Udomkesmalee and Steven C. Suddarth, *Vigilante: Ultrafast Smart Sensor for Target Recognition and Precision Tracking in a Simulated CMD Scenario* (Washington, DC: Ballistic Missile Defense Organization, 1997), 1.

Guna Seetharaman, an engineer: Guna Seetharaman, interview with the author, June 23, 2017.

115 *Vaidya's team built software that could:* Arnie Heller, "From Video to Knowledge," *Science & Technology Review,* April/May 2011, https://str.llnl.gov/AprMay11/vaidya.html, and James Poss, interview with the author, July 7, 2016.

generated such stable imagery: John Marion, interview with the author, June 7, 2017.

a second motive for the Persistics stabilization: Sheila Vaidya, interview with the author, January 12, 2017.

the Persistics computer would simply isolate and track: Heller, "From Video to Knowledge," and Alan Aaron Bernstein, "Foveated Coding for Persistics" (master's thesis, University of Texas at Austin, December 2012), 1.

an analyst would simply have to: Heller, "From Video to Knowledge," and Marion, interview, June 7, 2017.

The software could increase the video's resolution: Vaidya, interview, January 12, 2017, and Heller, "From Video to Knowledge."

116 *In 2011, when Gorgon Stare I was deployed . . . and in the field:* Vaidya, interview, May 17, 2018.

A Livermore spokesperson disputed: Steve Wampler, email to the author, April 25, 2018.

Though Vaidya's researchers were accustomed . . . "graduate school!": Vaidya, interview, January 12, 2017.

By the time Persistics was drawn down: Vaidya, interview, January 12, 2017.

117 *had no patience for the sophisticated analytical tools:* Cooter, interview, June 1, 2017.

Though Constant Hawk had a watchbox feature: Ross, interview, December 16, 2016.

there were at least 14 "state of the art": Rodney LaLonde, Dong Zhang, and Mubarak Shah, "ClusterNet: Detecting Small Objects in Large Scenes by Exploiting Spatio-Temporal Information," Center for Research in Computer Vision, University of Central Florida, December 4, 2017, available through ArXiv at https://arxiv.org/abs/1704.02694, p. 7.

the Air Force spent several million dollars integrating: Department of the Air Force, *Fiscal Year (FY) 2019 Budget Estimates,* Air Force Justification Book Volume 3b of 3, Research, Development, Test & Evaluation, Air Force Vol–III Part 2: "675291 / Gorgon Stare," February 2018, 20 of 34.

"Is that building a weapons cache?": Vaidya, interview, February 7, 2017.

A similar program developed by the Air Force Research Laboratory: Jianjun Gao et al., "Context-Aware Tracking with Wide-Area Motion Imagery," SPIE Newsroom, June 7, 2013, doi: 10.1117/2.1201305.004888.

programs like DARPA's Wide-Area Network Detection: John Keller, "DARPA Seeks to Develop Sensor Networks Able to Detect Threats in Densely Populated Cities," *Military & Aerospace Electronics,* May 5, 2010, www.militaryaero space.com/articles/2010/05/darpa-seeks-to-develop.html.

118 *In one particularly revealing research paper:* Reid Porter, Andrew M. Fraser, and Don Hush, "Narrowing the Semantic Gap in Wide Area Motion Imagery," *IEEE Signal Processing Magazine* 27, no. 5 (September 2010), doi: 10.1109/MSP.2010.937396, 56–65.

tracking software developed by the Australian firm: Sentient Vision, "ViDAR," www.sentientvision.com/products/vidar/.

The US Coast Guard is interested: Sentient Vision, "Sentient Demonstrates ViDAR Optical Radar to the US Coast Guard," press release, September 29, 2016, www.sentientvision.com/2016/09/29/sentient-demonstrates-vidar-opti cal-radar-us-coast-guard/.

Whenever a brute-force analyst: Interview with Mark Cooter, June 1, 2017.

the second main task in imagery analysis: Bianchetti and MacEachren, "Cognitive Themes Emerging from Air Photo Interpretation Texts Published to 1960," 565.

119 *During the Cold War . . . into the Barents Sea:* David Doyle, interview with the author, May 16, 2016.

120 *A senior analyst who worked:* Ingard Clausen and Edward A. Miller, *Intelligence Revolution 1960: Retrieving the Corona Imagery That Helped Win the Cold War* (Chantilly, VA: Center for the Study of National Reconnaissance, April 2012), 68.

A good analyst could look at a column of tanks: Patrick Eddington, interview with the author, June 28, 2016; and Patrick Eddington, *Long Strange Journey* (Shelbyville, KY: Wasteland Press, 2011).

8. Ghost in the Stare

121 *One day in the spring of 2008:* This account comes from John Montgomery, in an interview with the author on December 19, 2016.

they even committed to memory the exact spots: Marc Schanz, "The Indispensable Weapon," *Air Force Magazine,* February 2010, 33–34.

122 *The US government, as a signatory:* US Department of State, Comprehensive Nuclear Test-Ban Treaty (CTBT), https://www.state.gov/t/avc/c42328.htm.

123 *Charles Law and Bill Hoffman, the company's founders:* Charles Law and Bill Hoffman, interview with the author, January 31, 2017.

Determined that their work on the concept: Law and Hoffman, interview, January 17, 2017.

"Well we don't want to start with the IED going off": John Marion, interview with the author, November 30, 2016.

The whole point of the vast: Rick Atkinson, "Left of Boom: The Fight Against Improvised Explosive Devices," *Washington Post,* September 30, 2007, available through Internet Archive at https://web.archive.org/web/2017

1231194909/http://www.washingtonpost.com/wp-dyn/content/article/2007/
09/28/AR2007092801888.html.

One day Hoogs was watching...before it happened?: Anthony Hoogs, interview with the author, January 31, 2017.

124 *Known as activity-based intelligence:* Cathy Johnston, "(U) Modernizing Defense Intelligence: Object Based Production and Activity Based Intelligence," PowerPoint presentation, Defense Intelligence Agency Innovation Day, June 27, 2013, Washington, DC.

They blended seamlessly into the local population: Karen E. Thuermer, "Counter-IED Technologies Critical Worldwide," *Tactical ISR Technology* 4, no. 1 (February 2014): 10.

Donald Rumsfeld's widely pilloried remark: David A. Graham, "Rumsfeld's Knowns and Unknowns: The Intellectual History of a Quip," *Atlantic,* March 27, 2014, https://www.theatlantic.com/politics/archive/2014/03/rumsfelds-knowns-and-unknowns-the-intellectual-history-of-a-quip/359719/.

"like looking in a global ocean": Letitia Long, "Activity Based Intelligence: Understanding the Unknown," *Intelligencer: Journal of U.S. Intelligence Studies* 20, no. 2 (Fall/Winter 2013): 7.

125 *a report by a senior Pentagon intelligence official:* Bob Arbetter, Deputy Director, Development and Enabling (DE), Office of the Undersecretary of Defense (Intelligence), "Activity Based Intelligence and Human Dimension Analytics," cited in Edwin C. Tse, "IMSC Spring Retreat Activity Based Intelligence Challenges," PowerPoint presentation to the Integrated Media Systems Center Spring Retreat, University of Southern California, Davidson Conference Center, March 7, 2013.

"Bad guys do bad things": US Defense Advanced Research Projects Agency, "DARPA Advances Video Analysis Tools," posted on Phys.org, June 24, 2011, https://phys.org/news/2011-06-darpa-advances-video-analysis-tools.html#jCp.

The activity that Hoogs discovered: Hoogs, interview, January 31, 2017.

When Kitware, along with researchers: Ibid.

126 *engineers themselves didn't know the difference:* Ibid.

Kitware won an $11 million contract: Michael Peck, "U.S. Military Technology Projects Target Automated Imagery Analysis," Space News, November 29, 2010, http://spacenews.com/us-military-technology-projects-target-automated-imagery-analysis/.

This initiative enabled Kitware: Hoogs, interview, January 31, 2017.

127 *According to a document outlining:* Information Processing Techniques Office, Defense Advanced Research Projects Agency, *DARPA-BAA-09-55 Persistent Stare Exploitation and Analysis System (PerSEAS) Broad Agency Announcement (BAA)* (Arlington, VA: DARPA, 2009), 6, 8–10.

It didn't matter if the computer couldn't catch everything: Hoogs, interview, January 31, 2017.

In 2010 DARPA awarded Kitware: Kitware, "DARPA Awards Kitware a $13.8 Million Contract for Online Threat Detection and Forensic Analysis in Wide-Area Motion Imagery," press release, July 19, 2010, https://blog.kitware.com/darpa-awards-kitware-a-13-8-million-contract-for-online-threat-detection-and-forensic-analysis-in-wide-area-motion-imagery/.

A single data set might include 10 million: Hoogs, interview, January 31, 2017.

129 *analysts began to notice that if targets:* Marion, interview, June 7, 2017.

As early as 2005, engineers: Ibid.

As a solution, the PerSEAS engineers: Ibid.

a clear indicator, according to John Marion: Marion, interview, June 7, 2017.

Hoogs pointed out that while: Hoogs, interview, January 31, 2017.

130 *At GE Global Research, Hoogs had worked on a system:* "U.S. Navy's PANDA Technology to Detect 'Deviant' Ships," Homeland Security News Wire, November 13, 2009, www.homelandsecuritynewswire.com/us-navys-panda-technology-detect-deviant-ships, and interview with Anthony Hoogs, January 31, 2017.

Credit card companies employ a similar strategy: David Pendall, "The Promise of Persistent Surveillance: What Are the Implications for the Common Operating Picture?" Monograph AY 04-05, United States Army Command and General Staff College, School of Advanced Military Studies, Leavenworth, KS, May 26, 2005, 30.

Using techniques known as cluster analysis: Hoogs, interview, January 31, 2017.

Using similar techniques, the defense contractor Harris: Bernard V. Brower, Jason Baker, and Brian Wenink, "Wide-Area Motion Imagery for Multi-INT Situational Awareness," Harris report presented at NATO SET-241, May 26, 2017, Quebec, doi: 10.14339/STO-MP-SET-241, p. 3.

131 *From the sound of it, Hoogs said:* Hoogs, interview, January 31, 2017.

In 2014 the Air Force Research Laboratory described KWIVER: Air Force Research Laboratory, "Promoting Probabilistic Programming Systems (PPS) Development in Probabilistic Programming for Advancing Machine Learning (PPAML)," AFRL-RI-RS-TR-2018-073 (Rome, NY: Air Force Materiel Command, March 2018), p. 6, http://www.dtic.mil/dtic/tr/fulltext/u2/1050323.pdf.

It also built event-detection software: Hoogs, interview, June 23, 2017.

All told, Kitware has received almost $40 million: Michael Peck, "U.S. Military Technology Projects Target Automated Imagery Analysis," Space News, November 29, 2010, http://spacenews.com/us-military-technology-projects-target-automated-imagery-analysis/.

From 2009 to 2013 it consistently ranked: "Kitware," Inc., https://www.inc.com/profile/kitware.

132 *researchers . . . have suggested that a finely tuned:* Reid Porter, Andrew M. Fraser, and Don Hush, "Narrowing the Semantic Gap in Wide Area Motion Imagery," *IEEE Signal Processing Magazine,* 2010, 12.

Founded in 2004 by two engineers: Interview with Lawrence Carin in Kathleen Yount, "Signal Innovations Group," Duke University Electrical & Computer Engineering, https://ece.duke.edu/about/news/entrepreneurs/sig nal-innovations-group.

The company has worked with: "Signal Innovations Group: Innovative Technology to Help Interpret Complex Data," CTOvision.com, March 7, 2012, https://ctovision.com/signal-innovations-group-innovative-technology-to-help-interpret-complex-data/. As of August 2018, this page had been removed from the website.

A "Competition Sensitive" PowerPoint presentation: Signal Innovations Group, "WAMI for Multi-Source MOVINT," PowerPoint presentation, 2013, available online at https://pdfs.semanticscholar.org/presentation/f90e/4a8ad9 3d2248a631b86a6d78d1a2485ad53a.pdf.

133 *It appears that the company demonstrated:* Ibid.

The presentation appears to suggest: Ibid.

In addition to matching vehicle activity patterns: Ibid.

134 *In 2014 the defense contractor BAE:* BAE Systems, "BAE Systems Completes Acquisition of Signal Innovations Group," September 30, 2014, https://www.baesystems.com/en-us/article/bae-systems-completes-acquisition-of-signal-innovations-group.

BAE now sells software: Katherine Owens, "New BAE Systems Software Now Both Tracks and Interprets ISR Data," *Defense Systems,* June 22, 2017, https://defensesystems.com/articles/2017/06/22/bae-software.aspx.

the company appears to have used the city's medical industry: Signal Innovations Group, "WAMI for Multi-Source MOVINT."

In early 2017, after more than a decade: Cheryl Pellerin, "Project Maven to Deploy Computer Algorithms to War Zone by Year's End," US Department of Defense, DOD News, Defense Media Activity, July 21, 2017, https://www.de fense.gov/News/Article/Article/1254719/project-maven-to-deploy-computer-algorithms-to-war-zone-by-years-end/.

The goal of the program is simple: Tajha Chappellet-Lanier, "Pentagon's Project Maven Responds to Criticism: 'There Will Be Those Who Will Partner with Us,'" *FedScoop,* May 1, 2018, https://www.fedscoop.com/project-maven-artificial-intelligence-google/.

the program's first experiment: Gregory C. Allen, "Project Maven Brings AI to the Fight Against ISIS," *Bulletin of the Atomic Scientists,* December 21, 2017, https://thebulletin.org/project-maven-brings-ai-fight-against-isis11374.

135 *Among the software's many features:* Jack Shanahan, "Disruption in UAS: The Algorithmic Warfare Cross-Functional Team (Project Maven)," Power-

Point presentation, Office of the Secretary of Defense, Under Secretary of Defense for Intelligence, Washington, DC, October 26, 2017.

One of the contractors on the project: Matthew Zeiler, "Why We're Part of Project Maven," Clarifai, June 13, 2018, https://www.clarifai.com/blog/why-were-a-part-of-project-maven.

software capable of analyzing: Clarifai, "Models," https://clarifai.com/models.

By the end of 2018 the program will have deployed: Shanahan, "Algorithmic Warfare Cross-Functional Team (AWCFT) aka Project Maven."

136 *but a Pentagon budget document released in early 2018 noted:* US Department of Defense, "Reprogramming Action — Internal: Intelligence Surveillance and Reconnaissance Request," July 10, 2018, p. 3.

In a presentation to members of the Royal Australian Air Force: Shanahan, "Disruption in UAS."

In its budget request for 2019: Dan Gettinger, "Summary of Drone Spending in the FY 2019 Defense Budget Request," Center for the Study of the Drone, April 2018, p. 9, https://dronecenter.bard.edu/files/2018/04/CSD-Drone-Spending-FY19-Web-1.pdf.

Partners in the initiative include: Shanahan, "Algorithmic Warfare Cross-Functional Team (AWCFT) aka Project Maven."

DARPA Mind's Eye program: Bruce A. Draper, "Mind's Eye," Colorado State University, www.cs.colostate.edu/~draper/MindsEye.php.

"but," the researchers lamented: Ibid.

137 *YouTube's software has been trained on over 100 billion examples:* Paul Covington, Jay Adams, and Emre Sargin, "Deep Neural Networks for YouTube Recommendations," RecSys 2016, Boston, September 15–19, 2016.

It is for similar reasons that Facebook's: Naomi LaChance, "Facebook's Facial Recognition Software Is Different from the FBI's. Here's Why," NPR, May 18, 2016, https://www.npr.org/sections/alltechconsidered/2016/05/18/477819617/facebooks-facial-recognition-software-is-different-from-the-fbis-heres-why.

138 *"is nothing short of revolutionary": Perspectives on Research in Artificial Intelligence and Artificial General Intelligence Relevant to DoD,"* JSR-16-Task-003 (McLean, VA: MITRE Corporation, JASON Program Office, January 2017), 9.

When paired with a deep-learning system trained on ImageNet: Ross, interview, December 20, 2016.

A deep-learning-based tracking and detection system: Rodney LaLonde, Dong Zhang, and Mubarak Shah, "ClusterNet: Detecting Small Objects in Large Scenes by Exploiting Spatio-Temporal Information," December 4, 2017, arXiv: 1704.02694v2, 7.

Much of Project Maven's early success: Shanahan, "Algorithmic Warfare Cross-Functional Team (AWCFT) aka Project Maven."

Under Project Maven, when the system misidentifies: Lara Seligman, "How

Shadowy 'Project Maven' Uses AI to Mine Combat Data," *Aviation Week & Space Technology,* May 7, 2018, aviationweek.com/program-management-cor ner/how-shadowy-project-maven-uses-ai-mine-combat-data.

139 *"A pilot trained in Philadelphia":* Vijayan Asari, interview with the author, March 14, 2017.

If an operation moves from: Zhicong Qiu et al., "Actively Learning to Distinguish Suspicious from Innocuous Anomalies in a Batch of Vehicle Tracks," *SPIE Proceedings* 9079, Ground/Air Multisensor Interoperability, Integration, and Networking for Persistent ISR V, 90790G (2014), doi: 10.117/12.2052778.

In one example provided by: Shanahan, "Algorithmic Warfare Cross-Functional Team (AWCFT) aka Project Maven."

Montgomery had watched a YouTube video: John Montgomery, interview, December 19, 2016.

Much of the team from Lawrence Livermore's: Sheila Vaidya, interview with the author, January 12, 2017.

140 *In the fall of 2017 the Office of the Secretary of Defense selected Google:* Scott Shane, Cade Metz, and Daisuke Wakabayashi, "How a Pentagon Contract Became an Identity Crisis for Google," *New York Times,* May 30, 2018, https://www.nytimes.com/2018/05/30/technology/google-project-maven-pentagon.html.

In around 2013 Google signed: Department of Defense, Research and Engineering Enterprise, "DoD Scientist of the Quarter: Mr. Daniel Uppenkamp, Air Force Research Laboratory (AFRL)," March 28, 2014, https://www.acq.osd.mil/chieftechnologist/scientist/2014-4thQtr.html.

A CRADA is a partnership: Shane Harris, *@War: The Rise of the Military-Internet Complex* (Boston: Houghton Mifflin Harcourt, 2014), 175.

As a result of the AFRL-Google project: Department of Defense, Research and Engineering Enterprise, "DoD Scientist of the Quarter: Mr. Daniel Uppenkamp."

141 *In its 2019 budget request, the Special Operations Command: Fiscal Year (FY) 2019 President's Budget Operation and Maintenance, Defense-Wide,* United States Special Operations Command, February 2018, Comms-817.

The Air Force spokesperson would not confirm: Kenneth Schultz, email to the author, May 16, 2018.

"couldn't be assured that it would align with our AI Principles": "Google Drops Out of Bidding for Massive Pentagon Cloud Contract," Agence France-Presse, October 9, 2018, https://www.voanews.com/a/google-drops-out-bid ding-massive-pentagon-cloud-contract/4605813.html.

"if big tech companies are going to turn their back on the US Department of Defense, this country is going to be in trouble": Dave Lee, "Amazon's Bezos: US Needs to Be Defended," BBC News, October 15, 2018, https://www.bbc.com/news/technology-45871248.

All 17 intelligence agencies: Frank Konkel, "The Details About the CIA's Deal with Amazon," *Atlantic,* July 17, 2014, https://www.theatlantic.com/technology/archive/2014/07/the-details-about-the-cias-deal-with-amazon/374632/.

142 *"We are not going to withdraw from the future":* Brad Smith, "Technology and the US Military," Microsoft on the Issues, October 26, 2018, https://blogs.microsoft.com/on-the-issues/2018/10/26/technology-and-the-us-military/.

"Finally: An App That Can Identify the Animal You Saw on Your Hike": Ed Young, "Finally: An App That Can Identify the Animal You Saw on Your Hike," *Atlantic,* July 27, 2017, https://www.theatlantic.com/science/archive/2017/07/an-app-for-identifying-animals-and-plants/535014/.

"Google Uses AI to Find Your Fine-Art Doppelgänger": Stephanie Mlot, "Google Uses AI to Find Your Fine-Art Doppelgänger," Geek.com, January 16, 2018, https://www.geek.com/tech/google-uses-ai-to-find-your-fine-art-doppel ganger-1727948/.

9. New Dimensions

143 *The first truly all-seeing CCTV:* William Ross, interview with the author, December 20, 2016.

Operators could pan, tilt, and zoom . . . 30 days: Department of Homeland Security, "Imaging System for Immersive Surveillance: New Video Camera Sees It All," August 24, 2015, https://www.dhs.gov/imaging-system-immer sive-surveillance-new-video-camera-sees-it-all.

"You don't miss anything": Ross, interview, December 20, 2016.

After the initial tests at Logan: Ross and Kelly, "Wide-Area Motion Imaging (WAMI) Technology and Systems."

In 2016 MIT transferred: Crawford, interview, February 7, 2017.

144 *In 2018 the company installed a unit:* Crawford, interview, March 1, 2018.

Soon after beginning work on . . . war zones: MIT Lincoln Laboratory, *2013 Annual Report,* 22, https://www.ll.mit.edu/publications/Annual_Report_2013. pdf.

In early 2017 Customs and Border Protection outlined: Spencer Woodman, "U.S. Seeks to Double Surveillance Towers Along the Mexican Border," Intercept, January 2017, https://theintercept.com/2017/01/27/u-s-seeks-to-double-video-surveillance-towers-along-mexican-border/.

According to a solicitation for proposals: US Customs and Border Protection, "RVSS Operational Requirements," solicitation document fragment, https://foi arr.cbp.gov/docs/Border_Wall_Records/2018/1021114406_2142/1805311017_ BW_FOIA_CBP_001972___001979.pdf1.

145 *The Israeli firm Elbit Systems:* Elbit Systems of America, "Elbit Systems of America Showcases Border Solutions Expertise at Border Security Expo," press release, April 6, 2017, https://www.marketwatch.com/press-release/

elbit-systems-of-america-showcases-border-solutions-expertise-at-border-security-expo-2017-04-06.

surveillance programs in "populated areas": Elbit Systems, "Company Profile: Next Is Now," http://elbitsystems.com/media/NIN_2017.pdf, p. 61.

Soccer fans at the Corinthians Arena in São Paulo: Hikvision, "Hikvision Announces Partnership with Sao-Paolo Based Corinthians," press release, March 23, 2018, markets.businessinsider.com/news/stocks/hikvision-announces-partnership-with-sao-paolo-based-corinthians-1019136412.

Logipix, a Hungarian company: Logipix, "LOGIPIX Panorama Cameras," www.logipix.com/index.php/components/logipix-hardware-components/panoramic-cameras.

The startup Aqueti sells a gigapixel camera: David J. Brady, email to the author, January 31, 2018.

The 2017 Little League World Series: PR Newswire, "Axis Communications, Extreme, Lenel and Milestone Team Up to Secure 2017 Little League Baseball® World Series," press release, August 24, 2017, markets.businessinsider.com/news/stocks/Axis-Communications-Extreme-Lenel-and-Milestone-Team-up-to-Secure-2017-Little-League-Baseball-World-Series-1002282579.

Axis Communications, the company that sells the 20-megapixel camera: Axis Communications, "Axis Camera Reads Distant License Plate," YouTube video, 1:19, February 12, 2014, https://www.youtube.com/watch?v=2mPuUIBIagQ.

Thanks to add-on facial-recognition and activity-detection software: Crawford, interview, March 1, 2018.

The cameras at the Corinthians stadium: Hikvision, "Hikvision Announces Partnership with Sao-Paolo Based Corinthians."

146 *Following terrorist attacks like the 2017 bombing of an Ariana Grande concert:* Hershman, keynote speech, and Sean Morrison, "Manchester Arena Attack: Police Plan to Put Suicide Bomber Salman Abedi's Brother on Trial," *Evening Standard,* May 16, 2018, https://www.standard.co.uk/news/crime/manchester-arena-attack-police-plan-to-bring-suicide-bomber-salman-abedis-brother-to-trial-a3840181.html.

Since 2016 Amazon has been selling facial- and object-recognition: Amazon Web Services, "Amazon Rekognition," https://aws.amazon.com/rekognition/.

The Orlando Police Department and the Washington: Nick Wingfield, "Amazon Pushes Facial Recognition to Police. Critics See Surveillance Risk," *New York Times,* May 22, 2018, https://www.nytimes.com/2018/05/22/technology/amazon-facial-recognition.html.

cities like Johannesburg: Muhammad Hussain, "Will Plugging into 'Advanced Monitoring' Allow SA to Reclaim the Streets?," *City Press,* May 26, 2018, https://city-press.news24.com/News/will-plugging-into-advanced-monitoring-allow-sa-to-reclaim-the-streets-20180525.

and Singapore: Aradhana Aravindan and John Geddie, "Singapore to Test Facial Recognition on Lampposts, Stoking Privacy Fears," Reuters, April 13, 2018, https://www.reuters.com/article/us-singapore-surveillance/singapore-to-test-facial-recognition-on-lampposts-stoking-privacy-fears-idUSKBN1H K0RV.

In Moscow, officials have suggested: Devin Coldewey, "Moscow Officially Turns on Facial Recognition for Its City-Wide Camera Network," TechCrunch, September 28, 2017, https://techcrunch.com/2017/09/28/moscow-officially-turns-on-facial-recognition-for-its-city-wide-camera-network/.

which has collaborated with the startup SeeQuestor: James Temperton, "One Nation Under CCTV: The Future of Automated Surveillance," *Wired,* August 17, 2015, www.wired.co.uk/article/one-nation-under-cctv.

IARPA, the US intelligence community's advanced-research laboratory, hopes: Terry Adams, "Deep Intermodal Video Analytics: Proposers' Day Brief," PowerPoint presentation, Intelligence Advanced Research Projects Activity, July 12, 2016.

147 *one could eventually use gigapixel cameras to broadcast sports games:* David J. Brady, "Gigapixel Television," presented at the 14th Takayanagi Kenjiro Memorial Symposium, Shizuoka University, Hamamatsu, Japan, November 27, 2012, available at http://www.disp.duke.edu/projects/AWARE/papers/TakayanagiProceedingsBrady.pdf; and David J. Brady et al., "Parallel Cameras," *Optica* 5 (2018): 127–37.

including software that turns aerial imagery of cities: Brian Leininger, Richard E. Nichols, and Casey Gragg, "UrbanScape," *SPIE Proceedings* 6578, Defense Transformation and Net-Centric Systems (2007), doi: 10.1117/12.724498; and Brian Leininger, interview with the author, April 7, 2017.

The Duke team has already demonstrated . . . The Matrix: Matt Hartigan, "Connected Gloves and 'Bullet Time': NBC Thinks Technology Can Make Boxing Cool," *Fast Company,* March 6, 2015, https://www.fastcompany.com/3042958/nbc-primetime-boxing-connected-gloves-high-tech-cameras.

148 *"All that we've learned from WAMI in air":* Sheila Vaidya, interview with the author, February 13, 2017.

While he was developing ARGUS at DARPA, Brian Leininger: Leininger, interview, April 7, 2017.

In 2013 the SkySat-1, a satellite about the size of a dishwasher: Mario Dubois, "Skysat: A New Generation of HD Imaging Satellites," Substance ÉTS, February 4, 2014, substance-en.etsmtl.ca/skysat-a-new-generation-of-hd-imaging-satellites; and Terra Bella, "Skybox Imaging Captures World's First High-Resolution, HD Video of Earth from Space (1080p HD)," YouTube video, 1:19, December 23, 2013, https://www.youtube.com/watch?v=fCrB1t8MncY.

A number of groups are working, quietly: Arslan Basharat, "Adapting Video and Image ISR to A2/AD," PowerPoint presentation, Next Generation Intelligence Surveillance and Reconnaissance, June 24–25, 2014, Arlington, VA.

149 *A single imaging satellite built by Digital Globe:* Jonathan D. Woods, "Watch a High-Tech Satellite Get Launched into Space," *Time,* August 13, 2014, time.com/3106769/satellite-digitalglobe-space-technology/.

Most US government satellites cost well over $1 billion: Marshall Curtis Erwin, *Intelligence, Surveillance, and Reconnaissance (ISR) Acquisition: Issues for Congress* (Washington, DC: Congressional Research Service, April 16, 2013), 6.

cost as little as $40,000 to build: Jos Heyman, "Focus: CubeSats — A Costing + Pricing Challenge," *SatMagazine,* October 2009, www.satmagazine. com/story.php?number=602922274.

and about $80,000 to put into orbit: Charles L. Gustafson and Siegfried W. Janson, "Think Big, Fly Small," *Crosslink,* Summer 2014, p. 1, www.aero space.org/crosslinkmag/fall-2014/think-big-fly-small/.

The contractor General Atomics has proposed: Caleb Henry, "General Atomics Ramping Cubesat Production, Muses Railgun Smallsat Launcher," SpaceNews, October 12, 2017, spacenews.com/general-atomics-ramping-cubesat-production-muses-railgun-smallsat-launcher/.

150 *The Air Force:* Debra Werner, "Air Force to Bolster Weather Capabilities with Small Satellites and Sensors," SpaceNews, January 11, 2018, spacenews. com/air-force-to-bolster-weather-capabilities-with-small-satellites-and-sen sors/.

the Navy: Kyra Wiens, "With SpinSat Mission, NRL Will Spin Small Satellite in Space with New Thruster Technology," US Naval Research Laboratory, September 18, 2014, https://www.nrl.navy.mil/media/news-releases/2014/with-spinsat-mission-nrl-will-spin-small-satellite-in-space-with-new-thruster-technology.

the Army: Sandra Erwin, "Army Space Project a Now-or-Never Moment for Low-Cost Military Satellites," SpaceNews, October 25, 2017, spacenews. com/army-space-project-a-now-or-never-moment-for-low-cost-military-sat ellites/.

and even the National Reconnaissance Office: "Colony-1 CubeSats of NRO," Earth Observation Portal, Directory, https://directory.eoportal.org/web/ eoportal/satellite-missions/c-missions/colony-1.

Special Operations Command runs a fleet: Noah Shachtman, "With New Mini-Satellites, Special Ops Takes Its Manhunts into Space," *Wired,* May 21, 2013, https://www.wired.com/2013/05/special-ops-mini-sats-manhunts/.

Under a contingency plan revealed in 2017: David E. Sanger and William J. Broad, "Tiny Satellites from Silicon Valley May Help Track North Korea Missiles," *New York Times,* July 6, 2017, https://www.nytimes.com/2017/07/06/ world/asia/pentagon-spy-satellites-north-korea-missiles.html?emc=edit_ nn_20170707&nl=morning-briefing&nlid=73250843&te=1.

Some in the Pentagon envision that eventually every analyst: Defense Advanced Research Projects Agency, "Harnessing Commercially Available Geospatial Imagery for Defense Analysis," October 11, 2017, https://www.darpa.mil/news-events/2017-10-11.

DARPA is working to build constellations: Todd Master, "Space Enabled Effects for Military Engagements (SeeMe)," Defense Advanced Research Projects Agency, www.darpa.mil/program/space-enabled-effects-for-military-engagements.

satellites weighing around 600 pounds that can be launched: Fiscal Year (FY) 2019 Budget Estimate, Defense-Wide Justification Book, Volume 1 of 5 Research, Development, Test & Evaluation: PE 0603287E: SPACE PROGRAMS AND TECHNOLOGY, U.S. Department of Defense, Defense Advanced Research Projects Agency, February 2018, 7 of 9.

In 2017 a single Russian launch rocket carried: "Russia Launches 73 Satellites into Orbit," Phys.org, July 14, 2017, https://phys.org/news/2017-07-russia-satellites-orbit.html.

while India's space agency is developing: D. S. Madhumathi, "ISRO Developing a Compact Launcher for Small Satellites," *Hindu,* December 11, 2017, www.thehindu.com/sci-tech/science/isro-developing-a-compact-launcher-for-small-satellites/article21420644.ece.

China's largest missile and rocket manufacturer: Zhao Lei, "Missile Giant Targets 20% of Market to Launch Small Satellites," *China Daily,* November 29, 2016, www.chinadaily.com.cn/china/2016-11/29/content_27509808.htm.

With its constellation of about 200 satellites: Ashlee Vance, "The Tiny Satellites Ushering in the New Space Revolution," *Bloomberg Businessweek,* June 29, 2017, https://www.bloomberg.com/news/features/2017-06-29/the-tiny-satellites-ushering-in-the-new-space-revolution.

Capella Space, which plans to use a constellation: Tim Fernholz, "There's One Big Problem with Satellite Imagery, and a Space Startup Has Found a Solution for It," Quartz, January 7, 2017, https://qz.com/879270/theres-one-big-problem-with-satellite-imagery-and-a-space-startup-has-found-a-solution-for-it/.

151 *In 2016 the National Geospatial-Intelligence Agency awarded Planet:* National Geospatial-Intelligence Agency, "NGA Introductory Contract with Planet to Utilize Small Satellite Imagery," October 24, 2016, https://www.nga.mil/MediaRoom/PressReleases/Pages/NGA-introductory-contract-with-Planet-to-utilize-small-satellite-imagery.aspx.

Within a month of the program's launch: Kristin Quinn, "USGIF Hosts Second Small Satellite Workshop at NGA," *Trajectory Magazine,* November 15, 2016, trajectorymagazine.com/usgif-hosts-second-small-satellite-workshop-nga/.

Eventually, the technology and the rate of launches: James Crawford, CEO, Orbital Insight, interview with the author, March 15, 2017.

Another small-sat startup, BlackSky: John K. Hornsby, BlackSky Global, "A Revolutionary Change in Earth Observation," PowerPoint presentation, GEOSmart Asia, October 1, 2015, Kuala Lumpur, Malaysia.

152 *Charles Norton, an engineer who leads a team:* Charles Norton, interview with the author, March 22, 2017.

But in 2013 the Aerospace Corporation: Richard P. Welle and David Hinkley, "The Aerospace Nano/PicoSatellite Program," presentation at In-Space Non-Destructive Inspection Technology Workshop, July 15–16, 2014, Johnson Space Center, Houston, TX.

a camera called SPIDER: Lockheed Martin, "How We're Shrinking the Telescope: An Up-Close Look at SPIDER," January 19, 2016, https://www.lockheedmartin.com/us/innovations/011916-webt-spider.html.

A single array capable of creating a high-resolution: Patrick Tucker, "Future Spy Satellites Just Got Exponentially Smaller," Defense One, September 1, 2017, www.defenseone.com/technology/2017/09/future-spy-satellites-just-got-exponentially-smaller/140700/?oref=search_Spy%20satellites%20cheaper.

153 *In one test flight, the Zephyr S:* Angus Batey, "Record Flight Showcases Zephyr Pseudosatellite Capabilities," *Aviation Week & Space Technology,* August 14, 2018, aviationweek.com/technology/record-flight-showcases-zephyr-pseudosatellite-capabilities.

Orbital Insight, a San Francisco–based startup: Crawford, interview with the author, March 15, 2017.

Orbital provides services: Crawford, interview with the author, March 15, 2017.

154 *The CIA and NGA are working to develop:* Kevin McCaney, "NGA Joins SpaceNet Satellite Imagery Initiative," GCN, September 20, 2016, https://gcn.com/articles/2016/09/20/nga-spacenet.aspx.

Meanwhile, the Defense Innovation Unit: Tom Simonite, "The Pentagon Wants Your Help Analyzing Satellite Images," *Wired,* February 21, 2018, https://www.wired.com/story/the-pentagon-wants-your-help-analyzing-satellite-images/.

Work is also already underway to do the same for satellite video: Anthony Hoogs, interview with the author, January 31, 2017.

"start to get at wide-area motion imaging writ large": William Ross, interview with the author, December 20, 2016.

"real-time, continuous video of almost anywhere on Earth": EarthNow, "EarthNow to Deliver Real-Time Video via Large Satellite Constellation," press release, April 18, 2018, http://10z325bj2404dqj6e3lhft8y-wpengine.netdna-ssl.com/wp-content/uploads/2018/04/Press_Release_18April2018.pdf.

155 *is backed by a number of high-profile investors:* Ibid.

Advanced fighter jets: Lara Seligman, "Super Hornet Demonstrates 'Eye-Watering' Sensor Fusion," *Aerospace Daily & Defense Report,* May 24, 2018, aviationweek.com/defense/super-hornet-demonstrates-eye-watering-sensor-fusion.

164 *In 2017, after at least half a decade of development:* US Department of Defense, "Contracts," September 20, 2017, Release No: CR-183-17, https://www.defense.gov/News/Contracts/Contract-View/Article/1318702/. BAE's role is confirmed in Department of Defense *Fiscal Year (FY) 2018 Budget Estimates,* Air Force Justification Book: Research, Development, Test & Evaluation, Air Force Vol–II, May 2017, page 7.

Meanwhile, a tool developed under DARPA's: Fiscal Year (FY) 2015 Budget Estimates: Defense Advanced Research Projects Agency Defense Wide Justification Book Volume 1 of 5 Research, Development, Test & Evaluation, Defense-Wide, US Department of Defense, March 2014, 293.

NGA's Map of the World: Greg Slabodkin, "NGA's Map to Put a World of Geospatial Intell in One Place," Defense Systems, August 25, 2014, https://defensesystems.com/articles/2014/08/25/nga-map-of-the-world-geoint.aspx.

The agency's Multi-INT Analysis and Archive System and another program called QuellFire: General Dynamics Advanced Information Systems, *"MAAS® Scalable Software Suite for Motion Imagery (MI) and Intelligence Exploitation, Dissemination and Archiving,"* DS4-0614-2b, June 2015; and Cathy Johnston, "(U) Modernizing Defense Intelligence: Object Based Production and Activity Based Intelligence," PowerPoint presentation, Defense Intelligence Agency, June 27, 2013.

By 2020 the NGA predicts that any analyst: National Geospatial-Intelligence Agency, *2020 Analysis Technology Plan* (Springfield, VA: National Geospatial-Intelligence Agency, August 2014), 8.

BAE sells a software suite: Geoff Fein, "BAE Systems Adds Movement Tracking Capability to GXP Software," *Jane's International Defence Review,* June 8, 2017, www.janes.com/article/71241/bae-systems-adds-movement-tracking-capability-to-gxp-software.

the company had received over $400 million: Slabodkin, "NGA's Map to Put a World of Geospatial Intell in One Place," Defense Systems, August 25, 2014, https://defensesystems.com/articles/2014/08/25/nga-map-of-the-world-geoint.aspx; and John Keller, "BAE Systems Names Industry Team to Help DARPA Unify Imaging and Military Intelligence Sensors," *Military & Aerospace,* January 16, 2014, www.militaryaerospace.com/articles/2014/01/bae-insight-team.html.

SRC, a New York–based research corporation: Mark Pomerleau, "As DoD Seeks to Unburden Analysts, Industry Ramps Up Investments," C4ISRNET, March 13, 2018, https://www.c4isrnet.com/c2-comms/2018/03/13/as-dod-seeks-to-unburden-analysts-industry-ramps-up-investments/.

Virginia-based defense firm Leidos: Leidos, *Advanced Analytics Suite,* pamphlet, available through the Internet Archive at https://web.archive.org/web/20170205093703/https://www.leidos.com/products/software/advanced-analytics.

The defense giant Lockheed Martin: Lockheed Martin, *Hydra Fusion Tools,*™ https://www.lockheedmartin.com/us/products/cdl-systems/hydra-fusion-tools.html.

165 *A license for BAE's Hydra:* BAE Systems, *Authorized Federal Acquisition Service, Information Technology Schedule Price List, General Purpose Commercial Technology, Equipment, Software and Services* (San Diego, CA: BAE Systems Information & Electronic Systems Integration, June 2016), 20.

Lockheed Martin's software, at just $9,000: Lockheed Martin, "Lockheed Martin CDL Systems Software Products Price List," https://www.lockheedmartin.com/content/dam/lockheed/data/ms2/documents/cdl-systems/LM-CDL-Systems-Software-Products-Price-List.pdf.

In the hypothetical antiradar mission . . . rules of engagement: DARPAtv, "Collaborative Operations in Denied Environment (CODE) Human-System Interface Test," YouTube video, 2:15, May 13, 2016, https://www.youtube.com/watch?v=o8AFuiO6ZSs&feature=youtu.be.

When one device has trouble identifying: Ibid.

166 *If another drone detects:* Defense Advanced Research Projects Agency, "CODE Makes Headway on Collaborative Autonomy for Unmanned Aerial Systems," January 8, 2018, https://www.darpa.mil/news-events/2018-01-08.

It could also speed up the pace of battlefield decisions: US Department of Defense, *Persistent Intelligence, Surveillance, and Reconnaissance: Joint Integrating Concept Version 1.0* (Washington, DC: Department of Defense, March 29, 2007), 11.

And because drone swarms of this kind: Department of Defense, Defense Science Board, *The Role of Autonomy in DoD Systems* (Washington, DC: Office of the Under Secretary of Defense for Acquisition, Technology and Logistics, 2012), 45.

A second, more nightmarish: Timothy H. Chung, "OFFensive Swarm-Enabled Tactics (OFFSET)," PowerPoint presentation, Defense Advanced Research Projects Agency, n.d.

A program called Bedlam is a case in point: Randy Milbert, "Bedlam: One Raven UAV Tracking an Evasive Ground Vehicle," YouTube video, 5:49, January 6, 2012, https://www.youtube.com/watch?v=1OKNC4pg9BU; and Primordial, Inc., "Bedlam," www.primordial.com/index.php/projects/bedlam.

In 2017 researchers at Cornell announced a plan: Cornell Media Relations Office, "Did You Catch That? Robot's Speed of Light Communication Could Protect You from Danger," press release, April 10, 2017, mediarelations.cor

nell.edu/2017/04/10/did-you-catch-that-robots-speed-of-light-communica
tion-could-protect-you-from-danger/.

167 *About 80 local law enforcement agencies:* Mark Harris, "How Peter Thiel's
Secretive Data Company Pushed into Policing," *Wired,* August 9, 2017, https://
www.wired.com/story/how-peter-thiels-secretive-data-company-pushed-
into-policing/.
fusion centers modeled on military war rooms: Department of Homeland
Security, "Fusion Center Locations and Contact Information," https://www.
dhs.gov/fusion-center-locations-and-contact-information.
serve as a clearinghouse: US Department of Justice, Office of Justice Programs, Bureau of Justice Assistance, "Fusion Centers and Intelligence Sharing," n.d., https://it.ojp.gov/initiatives/fusion-centers.
According to guidelines drafted: US Department of Homeland Security, *The
Role of Fusion Centers in Countering Violent Extremism: Overview,* October
2012, https://it.ojp.gov/documents/roleoffusioncentersincounteringviolentex
tremism_compliant.pdf.

168 *Meanwhile, a number of US law enforcement departments:* Ali Winston,
"Palantir Has Secretly Been Using New Orleans to Test Its Predictive Polic-
ing Technology," Verge, February 27, 2018, https://www.theverge.com/2018
/2/27/17054740/palantir-predictive-policing-tool-new-orleans-nopd.
is rumored to have provided: Harris, "How Peter Thiel's Secretive Data
Company Pushed into Policing."
A representative I spoke with: Aaron Yue, Worldwide Public Sector Ser-
vices Solutions Architect, Microsoft, interview with the author, July 5, 2017.
Qatar's security agencies plan to use: Michael Hershman, keynote speech,
INTERPOL World 2017, July 5, 2017.

169 *In China, a company called Dahua Technology:* "In Your Face: China's All-
Seeing State," BBC News, December 10, 2017, www.bbc.com/news/av/world-
asia-china-42248056/in-your-face-china-s-all-seeing-state.
By some estimates, there will be 80 billion: Michael Kanellos, "152,000
Smart Devices Every Minute in 2025: IDC Outlines the Future of Smart
Things," *Forbes,* March 3, 2016, https://www.forbes.com/sites/michaelkane
llos/2016/03/03/152000-smart-devices-every-minute-in-2025-idc-outlines-
the-future-of-smart-things/#160423114b63.
"This immense, sparsely populated space of interconnected devices": Office
of the Under Secretary of Defense for Acquisition, Technology and Logistics,
Report of the Defense Science Board Summer Study on Autonomy (Washington,
DC: Department of Defense, June 2016), 88.

170 *"the greatest mass surveillance infrastructure ever conceived":* Julia Powles, "Internet of Things: The Greatest Mass Surveillance Infrastructure Ever?"
Guardian, July 15, 2015, https://www.theguardian.com/technology/2015/
jul/15/internet-of-things-mass-surveillance.

11. One Hell of a Fight

173   *an episode of the PBS documentary series* NOVA: "Rise of the Drones," *NOVA,*
PBS, January 23, 2013, www.pbs.org/video/nova-rise-drones-pro/.

Eighteen months earlier, a team of photographers in Vancouver: "Before the
Riot Version 1," Gigapixel.com, June 15, 2011, www.gigapixel.com/image/giga-
pan-canucks-g7.html.

"Even in our most pessimistic moments": Jay Stanley, "Drone 'Nightmare
Scenario' Now Has a Name: ARGUS," *Free Future* (blog), ACLU, February 21,
2013, https://www.aclu.org/blog/drone-nightmare-scenario-now-has-name-
argus.

174   *"We have a technology now to create the nightmare scenario":* Jay Stanley,
interview with the author, May 30, 2017.

Persistent Surveillance Systems had carried out: Persistent Surveillance
Systems, "2013 Aerial Surveillance Project," PowerPoint presentation, Dayton
City Commission, January 29, 2013, http://www.acluohio.org/wp-content/up
loads/2013/04/2013_0206AerialAirborneSurveillanceProgramPresentation
ToDaytonCityCommission.pdf.

When the proposed program was announced: ACLU Ohio, "Warrantless
Aerial Surveillance in Dayton," April 5, 2013, www.acluohio.org/archives/is
sue-information/warrantless-aerial-surveillance-in-dayton.

"In America," Stanley wrote in a post: Jay Stanley, "Ohio Aerial Surveillance System Moving Forward Without Having to Wait for FAA Drone Rules,"
Free Future (blog), ACLU, April 4, 2013, https://www.aclu.org/blog/national-
security/ohio-aerial-surveillance-system-moving-forward-without-having-
wait-faa-drone.

"While we believe there are real potential benefits": Jeremy P. Kelley, "City
Rejects Use of Surveillance Planes," *Dayton Daily News,* April 17, 2013, www.
daytondailynews.com/news/city-rejects-use-surveillance-planes/FiNRlRW0
gYI4hmYnZTMSiL/.

According to materials that McNutt showed me in Baltimore: Ross T. Mc-
Nutt, "Wide Area Surveillance in Support of Law Enforcement Privacy and
Policy Discussions," PowerPoint presentation, June 2014.

But the commissioner was clear: Kelley, "City Rejects Use of Surveillance
Planes."

175   *McNutt was, and remains:* Ross McNutt, interview with the author, July
14, 2016.

According to one of the people who participated: Kevin Rector and Luke
Broadwater, "Report of Secret Aerial Surveillance by Baltimore Police Prompts
Questions, Outrage," *Baltimore Sun,* August 24, 2016, www.baltimoresun.com/
news/maryland/baltimore-city/bs-md-ci-secret-surveillance-20160824-story.
html.

those limitations would be overcome: Jay Stanley, "Persistent Aerial Surveillance: Do We Want to Go There, America?," *Free Future* (blog), ACLU, February 7, 2014, https://www.aclu.org/blog/persistent-aerial-surveillance-do-we-want-go-there-america.

"You're asking, 'Why do we care about privacy?'": Stanley, interview, May 30, 2017.

176 *"Persistent surveillance allows users to track":* Matthew Feeney, "Baltimore Air Surveillance Should Cause Concerns," *The Hill,* August 25, 2016, thehill.com/blogs/pundits-blog/civil-rights/293329-baltimore-police-drones-should-cause-concerns.

If ARGUS were ever used by police: Jake Laperruque, "Preventing an Aerial Panopticon over American Cities," *Richmond Law Review* 51 (Spring 2017): 705–26, 713, 721.

"It's especially galling": Conor Friedersdorf, "Eyes over Compton: How Police Spied on a Whole City," *Atlantic,* April 21, 2014, https://www.theatlantic.com/national/archive/2014/04/sheriffs-deputy-compares-drone-surveillance-of-compton-to-big-brother/360954/.

"With the amount of technology out in today's age": Quoted in Amanda Pike and G. W. Schulz, "Hollywood-Style Surveillance Technology Inches Closer to Reality," Reveal, Center for Investigative Reporting, April 11, 2014, https://www.revealnews.org/article-legacy/hollywood-style-surveillance-technology-inches-closer-to-reality/.

"do we want to allow the government . . . is so wide": Stanley, interview, May 30, 2017.

177 *Stanley's position on surveillance technologies:* Ibid.

will hang up his hat in what he calls "his journey": Nathan Crawford, interview with the author, January 26, 2017.

One Persistent Surveillance Systems document claims: Ross McNutt, "Wide Area Surveillance and Counter Narco-Terrorism Operations," PowerPoint presentation, March 6, 2012, downloaded from https://www.slideshare.net/Shadowairs/waass-pss-counter-narco-terrorism-briefing.

178 *"But it wasn't a big leap for people to think":* William Ross, interview with the author, December 20, 2016.

Crawford said that during the brief collaboration: Crawford, interview, January 26, 2017.

"if it's not truly a capability": Crawford, interview, February 7, 2017.

"Now, there will always": Michael Meermans, interview, January 18, 2017.

"We all love our freedom": Yiannis Antoniades, interview, March 14, 2017.

179 *When Crawford gives demonstrations:* Crawford, interview, February 7, 2017.

In 2016 William Ray Walters: Joshua Barrie, "Man Catches 'Cheating' Wife by Filming Her Using Secret Drone," *Mirror,* November 15, 2016, www.mirror.co.uk/news/weird-news/man-catches-cheating-wife-filming-9261827.

"one hundred percent sure that I was going to kill that man": "Husband Who Used Drone to Catch Wife Cheating Confesses Homicidal Thoughts," *Inside Edition,* YouTube video, 1:52, November 16, 2016, https://www.youtube.com/watch?v=RAf0vX0fD2A.

Walters announced that he and Donna: YAOG, "Reconciliation," YouTube video, 1:52, December 6, 2017, https://www.youtube.com/watch?v=zCNbrTI3Sx4&t=1s.

In a video responding to questions: YAOG, "Q&A | We answered your questions," YouTube video, 19:46, December 23, 2017, https://www.youtube.com/watch?v=wnvHxaU-VkE&t=67s.

180 *"follow politicians from their houses to the gay bars and blackmail them":* Ross McNutt, interview with the author, July 14, 2016.

Walmart, he said: McNutt, interview with the author, July 14, 2016.

"a stalker's dream": Steven Suddarth, interview with the author, June 9, 2017.

"Sometimes people think, 'Hey'": McNutt, interview with the author, July 14, 2016.

"The fact of the matter is": Meermans, interview, January 18, 2017.

"I have nothing to hide": Richard Nichols, interview with the author, January 24, 2017.

181 *"The fact that you are making a call to your grandmother":* Sheila Vaidya, interview with the author, May 9, 2018.

In an earlier discussion, she said that she would rather: Vaidya, interview, March 2, 2017.

Anthony Hoogs, the Kitware engineer who is developing behavior-detection systems: Anthony Hoogs, interview with the author, January 31, 2017.

Bill Hoffman . . . noticed large groups of devices: Bill Hoffman, interview with the author, January 31, 2017.

Paul Boxer, the founder of Sentient Vision . . . wagers: Paul Boxer, interview with the author, September 8, 2016.

182 *"There has to be a discussion":* Ross, interview, December 20, 2016.

"It can't be, 'Oh, we're never going to do [WAMI]'": Ross, interview, December 20, 2016.

"come up with legislation about the uses of this information": Antoniades, interview, March 14, 2017.

Meermans said that existing US laws: Meermans, interview, January 18, 2017.

Crawford would only say that law enforcement: Crawford, interview, January 26, 2017.

"Because we're the contractor": In discussion with author, February 10, 2017.

183 *Marion did say that he "strongly disagreed":* John Marion, interview, February 10, 2017.

"We think we could have a pretty good value proposition": McNutt, interview, July 14, 2016.

To wit: Ibid.

"I would like to invite all of the newspapers": Ibid.

12. A Murder in Baltimore, Redux

184 *It would later emerge that McNutt had been emailing:* Brandon Soderberg, "Persistent Transparency: Baltimore Surveillance Plane Documents Reveal Ignored Pleas to Go Public, Who Knew About the Program, and Differing Opinions on Privacy," *City Paper,* November 1, 2016, http://www.citypaper.com/news/mobtownbeat/bcp-110216-mobs-aerial-surveillance-20161101-story.html.

McNutt told me when I was in Baltimore: McNutt, interview, July 14, 2016.

Most city council members and state legislators only learned about the program: Kevin Rector, "Cummings: Commissioner Davis 'Apologized Profusely' for Not Disclosing Surveillance Program," *Baltimore Sun,* September 2, 2016, www.baltimoresun.com/news/maryland/crime/bs-md-ci-cummings-davis-meeting-20160902-story.html.

185 *Baltimore's mayor, Stephanie Rawlings-Blake:* Kevin Rector and Luke Broadwater, "Report of Secret Aerial Surveillance by Baltimore Police Prompts Questions, Outrage," *Baltimore Sun,* August 24, 2016, www.baltimoresun.com/news/maryland/baltimore-city/bs-md-ci-secret-surveillance-20160824-story.html

"It wasn't the right time to do it": Interview with Don Roby, April 14, 2017.

"I'm angry that I didn't know about it and we did it in secrecy": Quoted in Rector and Broadwater, "Report of Secret Aerial Surveillance by Baltimore Police Prompts Questions, Outrage."

"Widespread surveillance violates every citizen's right to privacy": Ibid.

Carl Cooper, a suspect in the shooting . . . two hours: Kevin Rector, "Court Documents in Two Cases That Relied on Secret Aerial Surveillance Never Mention It," *Baltimore Sun,* August 29, 2016, www.baltimoresun.com/news/maryland/investigations/bs-md-sun-investigates-surveillance-cases-2016 0829-story.html.

Jay Stanley pointed out that: Jay Stanley, "Baltimore Aerial Surveillance Program Retained Data Despite 45-Day Privacy Policy Limit," *Free Future* (blog), ACLU, October 25, 2016, https://www.aclu.org/blog/free-future/baltimore-aerial-surveillance-program-retained-data-despite-45-day-privacy-policy.

The ACLU of Maryland issued a statement: cited in WMAR Baltimore, "Baltimore Police Defend Aerial Surveillance Program," August 24, 2016, available through Internet Archive at www.abc2news.com/news/region/baltimore-city/secret-aerial-surveillance-program-by-baltimore-police-sparks-criticism.

Congressman Elijah Cummings: Rector, "Cummings: Commissioner Davis 'Apologized Profusely.'" www.baltimoresun.com/news/maryland/crime/bs-md-ci-cummings-davis-meeting-20160902-story.html.

186 *"We've seen what this kind of behavior leads to in the past":* Barbara Haddock Taylor, "Baltimore Residents React to News of a Secret Baltimore Police Surveillance Project," *Baltimore Sun*, video, August 25, 2016, http://www.bal timoresun.com/news/88717085-132.html.

"I think there's more crime in this city than the cops": Ibid., 6.

Councilman Scott noted that although: Rector and Broadwater, "Report of Secret Aerial Surveillance by Baltimore Police Prompts Questions, Outrage."

addressed the chamber on behalf of the program: Maryland General Assembly, Judiciary Committee, "Briefing: Cell Site Simulator Technology, Historical Location Information, and Aerial Surveillance by Police," October 25, 2016, http://mgahouse.maryland.gov/mga/play/56d1a821-845a-4385-8308-fa6fe72fa0e4/?catalog/03e481c7-8a42-4438-a7da-93ff74bdaa4c.

187 *Over the course of the program, there were 21,243 calls to 911:* Police Foundation, *A Review of the Baltimore Police Department's Use of Persistent Surveillance (Baltimore Community Support Program)* (Washington, DC: Police Foundation, January 30, 2017), 15.

All told, leads collected by PSS: Ibid., 6.

In the murder case of Robert McIntosh: Jeff Hager, "Baltimore Aerial Surveillance Program Leads to Murder Arrest," ABC 2 Baltimore, August 25, 2016, https://www.abc2news.com/news/crime-checker/baltimore-city-crime/baltimore-aerial-surveillance-leads-to-murder-arrest.

In the final two months of the program: Maryland General Assembly, Judiciary Committee, "Briefing: Cell Site Simulator Technology."

"If the technology is up at the time someone": Ibid.

188 *In an interview with the* City Paper *a few days after the hearing:* Brandon Soderberg, "Persistent Transparency: Baltimore Surveillance Plane Documents Reveal Ignored Pleas to Go Public, Who Knew About the Program, and Differing Opinions on Privacy," *City Paper,* November 1, 2016, http://www.citypaper.com/news/mobtownbeat/bcp-110216-mobs-aerial-surveillance-20161101-story.html.

Smith assured the panel that the technology: Maryland General Assembly, Judiciary Committee, "Briefing: Cell Site Simulator Technology."

When Representative Charles E. Sydnor asked: Ibid.

"This is public space": Ibid.

"Memorandum of Law in Support of Constitutionality of Wide Airborne Surveillance": Baltimore Police Department, "Memorandum of Law in Support of Constitutionality of Wide Airborne Surveillance," in *A Review of the Baltimore Police Department's Use of Persistent Surveillance (Baltimore Community Support Program)* (Washington, DC: PoliceFoundation, 2017), https://www.po licefoundation.org/wp-content/uploads/2017/02/PF-Review-of-BCSP-Final.pdf, pp. 39–40.

189 *Smith seemed confident that:* Maryland General Assembly, Judiciary Committee, "Briefing: Cell Site Simulator Technology."

"highly recommends that a rigorous evaluation": Police Foundation, *A Review of the Baltimore Police Department's Use of Persistent Surveillance (Baltimore Community Support Program)*, 4.

"The police do not always" . . . *"the hallmark of courageous leadership":* Ibid., 20–21.

190 *But after a scathing report:* General Assembly of Maryland, 2018 Regular Session, "Task Force to Study Law Enforcement Surveillance Technologies," HB0578, February 21, 2018, mgaleg.maryland.gov/webmga/frmMain.aspx?pid =billpage&stab=01&id=hb0578&tab=subject3&ys=2018rs.

"just a couple of reporters that were hyperventilating": McNutt, interview, December 5, 2016.

an online poll from Baltimore Business Journal: "Business Pulse Poll: Are You Comfortable with City Police's Surveillance Program?," *Baltimore Business Journal,* 2016, https://www.bizjournals.com/baltimore/pulse/poll/are-you-comfortable-with-city-polices-surveillance-program/20446022.

A similar poll from the Baltimore Sun: "BPD's Secret Surveillance [Poll]," *Baltimore Sun,* August 25, 2016, www.baltimoresun.com/news/opinion/bal-bpds-secret-surveillance-poll-20160825-htmlstory.html.

According to the operational plan drawn up by McNutt: McNutt, interview, May 29, 2018.

The city hoped to focus the operation: Douglas Hanks, "Police Surveillance Would Target Northside from the Sky," *Miami Herald,* June 9, 2017, www.miamiherald.com/news/local/community/miami-dade/article155212459.html.

191 *The Miami-Dade Police Department's hope:* Carlos A. Giménez, memorandum to Honorable Chairman Esteban Bovo Jr. and Members, Board of County Commissioners, "Resolution Ratifying the Application of the County Mayor for Grant Funds from the U.S. Department of Justice Technology Innovation for Public Safety Grant 2017," July 6, 2017.

McNutt claims that he urged officials: McNutt, interview, May 29, 2018.

In its memo to the board: Giménez memorandum, "Resolution Ratifying the Application of the County Mayor for Grant Funds."

"You have no expectation of privacy": Carlos A. Giménez, quoted in Douglas Hanks, "To Fight Crime, Miami-Dade May Deploy Blanket Surveillance from the Air," *Miami Herald,* June 1, 2017.

Miami mayor Tomás Regalado told a local blogger: Ladra, "Some Mayors Say 'Ay!' to County Eye in the Sky Spy Plane," *Political Cortadito,* June 5, 2017, www.politicalcortadito.com/2017/06/05/mayors-ay-spy-plane/.

the ACLU of Florida issued a statement: American Civil Liberties Union, "ACLU of Florida Statement on the Miami-Dade Police Department's Proposal for Aerial Surveillance Tools," June 2, 2017, https://aclufl.org/2017/06/02/aclu-of-florida-statement-on-the-miami-dade-police-departments-proposal-for-aerial-surveillance-tools/.

"We do have the expectation that our movements": Jerry Iannelli, "Activist

Groups, Local Mayors Blast MDPD's Plan to Spy on Dade with Planes," *Miami New Times,* June 6, 2017, www.miaminewtimes.com/news/activists-condemn-mdpd-wide-area-surveillance-airplane-program-over-miami-9397981.

192 *At this point, anything that's going to help get the killings down":* Douglas Hanks, "Police Surveillance Would Target Northside from the Sky," *Miami Herald,* June 9, 2017, www.miamiherald.com/news/local/community/miami-dade/article155212459.html.

"I hear the voices": Douglas Hanks, "Miami-Dade Police Dropping Plan for Blanket Surveillance from the Air," *Miami Herald,* June 13, 2017, www.miami herald.com/news/local/community/miami-dade/article155867419.html.

In the spring of 2017 he doubled down: McNutt, interview, May 29, 2018.

he attended one of McNutt's presentations: Luke Broadwater, "A Group Is Trying to Get the Grounded Baltimore Police Surveillance Airplane Flying Again. The Pitch: It Can Catch Corrupt Cops," *Baltimore Sun,* February 22, 2018,www.baltimoresun.com/news/maryland/baltimore-city/bs-md-ci-police-plane-20180220-story.html.

McNutt recalls that Williams had spent most of the briefing: McNutt, interview, May 29, 2018.

193 *it had occurred to McNutt that the data could be used to validate:* Ibid.

He told me that Persistent Surveillance Systems had already helped: Ibid.

the beginning of Williams's transfiguration: Archie Williams, interview with the author, May 24, 2018.

McNutt and members of Community With Solutions approached a number of city leaders: McNutt, interview, May 29, 2018.

According to the Baltimore Sun, *Pugh:* Broadwater, "A Group Is Trying to Get the Grounded Baltimore Police Surveillance Airplane Flying Again."

Community With Solutions set up dozens: McNutt, interview, May 29, 2018.

Williams told me that he had personally: Williams, interview, May 24, 2018.

According to the Baltimore Sun: Broadwater, "A Group Is Trying to Get the Grounded Baltimore Police Surveillance Airplane Flying Again."

194 *The company did, however, provide:* McNutt, interview, May 29, 2018.

McNutt also allowed Community With Solutions . . . car seat: Williams, interview, May 24, 2018.

When I asked Williams about the privacy concerns: Ibid.

"That saves two people": Ibid.

"No, keep me on record": Ibid.

195 *"I got to see what it was like and it pissed me off":* McNutt, interview, May 29, 2018.

He told me that he subsidized PSS with revenue: Ibid.

he was awarded a large grant: "Dayton Tech Firm Wins Funding for 'Air Uber' System," *Dayton Business Journal,* July 16, 2018, https://www.bizjour

nals.com/dayton/news/2018/07/16/dayton-tech-firm-wins-funding-for-air-uber-system.html.

"From a business standpoint, I ought to close": McNutt, interview, May 29, 2018.

13. Where Power Meets Fury

198 *"the instrument of permanent, exhaustive, omnipresent surveillance":* Michel Foucault, *Discipline & Punish: The Birth of the Prison* (New York: Vintage Books, 1995), 214.

199 *Shortly after the invention of wiretapping:* Meyer Berger, "Tapping the Wires," *New Yorker,* June 18, 1938, https://www.newyorker.com/magazine/1938/06/18/tapping-the-wires.

200 *"If you're insecure":* Steven Suddarth, interview with the author, June 9, 2017.

To illustrate his point, Suddarth offered: Ibid.

201 *Persistent Surveillance Systems "doesn't do protests":* Ross McNutt, interview with the author, July 13, 2016.

202 *At one point during the 2016 CSP pilot program:* McNutt, interview, July 14, 2016.

203 *"some of them will be relevant in the future when an event happens":* Quoted in Kristin Quinn, "A Better Toolbox," *Trajectory* magazine, November 21, 2012, trajectorymagazine.com/a-better-toolbox-2/.

"mosaic theory of intelligence": Richards I. Heuer, "Psychology of Intelligence Analysis," Center for the Study of Intelligence, Central Intelligence Agency, 1999, 62.

204 *"the impression that we can 'observe'":* US Department of Defense, Office of the Chairman of the Joint Chiefs of Staff, *Major Combat Operations-Joint Operating Concept [MCO-JOC]* (Washington, DC: US Department of Defense, September 2004), 17.

205 *The Vision Lab at the University of Dayton is using:* Vijayan Asari, interview with the author, March 14, 2017.

In a test in a South African wildlife park: Elizabeth Bondi, Fei Fang, Mark Hamilton, Debarun Kar, Donnabell Dmello, Jongmoo Choi, Robert Hannaford, Arvind Iyer, Lucas Joppa, Milind Tambe, and Ram Nevatia, "SPOT Poachers in Action: Augmenting Conservation Drones with Automatic Detection in Near Real Time," Association for the Advancement of Artificial Intelligence, 2018, 1.

Kitware, the computer-vision company, has retooled: Charles Law, interview with the author, January 31, 2017.

dozens of cities in the United States: National Institute of Justice, "Predictive Policing," June 9, 2014, https://www.nij.gov/topics/law-enforcement/strategies/predictive-policing/Pages/welcome.aspx.

A number of states and jurisdictions: Sam Corbett-Davies et al., "A Computer Program Used for Bail and Sentencing Decisions Was Labeled Biased Against Blacks. It's Actually Not That Clear," *Washington Post,* October 17, 2016, https://www.washingtonpost.com/news/monkey-cage/wp/2016/10/17/can-an-algorithm-be-racist-our-analysis-is-more-cautious-than-propublicas/.

207 *When Livermore's Persistics software tracked:* Sheila Vaidya, interview with the author, May 9, 2018.

In one screenshot of a Project Maven: Jack Shanahan, "Disruption in UAS: The Algorithmic Warfare Cross-Functional Team (Project Maven)," Power-Point presentation, Office of the Secretary of Defense, Under Secretary of Defense for Intelligence, March 20, 2018.

These scores might be based: Marina Altynova et al., *Analyst Performance Measures Volume II: Information Quality Tools for Persistent Surveillance Data Sets* (Air Force Materiel Command, Air Force Research Laboratory, 711th Human Performance Wing, Wright-Patterson AFB OH 45433, October 7, 2011), 49.

Google's online computer-vision service: Google Cloud Platform, "Cloud Vision API," https://cloud.google.com/vision/.

208 *See Questor's CCTV-analysis program rates:* James Temperton, "One Nation Under CCTV: The Future of Automated Surveillance," *Wired,* August 17, 2015, www.wired.co.uk/article/one-nation-under-cctv.

In a paper titled (rather unoriginally, I might add) "Eye in the Sky": Amarjot Singh, Devendra Patil, and SN Omkar, "Eye in the Sky: Real-time Drone Surveillance System (DSS) for Violent Individuals Identification Using ScatterNet Hybrid Deep Learning Network," June 3, 2018, available on ArXiv at https://arxiv.org/pdf/1806.00746.pdf.

209 *When a US guided missile cruiser:* Jeremy R. Hammond, "The 'Forgotten' US Shootdown of Iranian Airliner Flight 655," *Foreign Policy Journal,* July 3, 2017, https://www.foreignpolicyjournal.com/2017/07/03/the-forgotten-us-shootdown-of-iranian-airliner-flight-655/.

In 2003 a similar error: Staff and agencies, "'Glaring Failures' Caused US to Kill RAF Crew," *Guardian,* October 31, 2006, https://www.theguardian.com/uk/2006/oct/31/military.iraq.

"high-regret actions": Greg Zacharias, lecture, "Air Force Future Operating Concept: Implications for Autonomy," Mitchell Institute, Rosslyn, Virginia, May 17, 2016.

Once it is lost, this trust is hard to regain: Paul D. Scharre, "The Opportunity and Challenge of Autonomous Systems," in *Autonomous Systems: Issues for Defence Policymakers,* ed. Andrew P. Williams and Paul D. Scharre (Norfolk, VA: NATO, Capability Engineering and Innovation Division, October 2015), 4.

210 *When IBM's Watson, an artificial intelligence program, appeared on the*

quiz show Jeopardy!: Henry Lieberman, "Watson on *Jeopardy,* Part 3," *MIT Technology Review,* February 16, 2011, https://www.technologyreview.com/s/422763/watson-on-jeopardy-part-3/.

211 *make Black and Latino drivers far more likely:* Richard Winton, "Black and Latino Drivers Are Searched Based on Less Evidence and Are More Likely to Be Arrested, Stanford Researchers Find," *Los Angeles Times,* June 19, 2017, www.latimes.com/local/lanow/la-me-ln-stanford-minority-drive-disparties-20170619-story.html.

One reviewer praised it for its "excellent design": Joshua Goldman, "Nikon CoolPix S630 Review: Nikon CoolPix S630," CNET, March 19, 2009, https://www.cnet.com/products/nikon-coolpix-s630/review/.

212 *"Did someone blink?":* Joz Wong, "Racist Camera! No, I Did Not Blink . . . I'm Just Asian!," Flickr, May 10, 2009, https://www.flickr.com/photos/jozjoz joz/3529106844.

a "smart" HP webcam released the following year: wzamen01, "HP Computers Are Racist," YouTube video, 2:15, December 10, 2009, https://www.you tube.com/watch?v=t4DT3tQqgRM.

In 2017, the CEO of a company called FaceApp: "FaceApp Sorry for 'Racist' Filter That Lightens Skin to Make Users 'Hot,'" BBC Newsbeat, April 25, 2017, www.bbc.co.uk/newsbeat/article/39702143/faceapp-sorry-for-racist-filter-that-lightens-skin-to-make-users-hot.

An FBI-sanctioned study from 2012 found that facial-recognition systems: Brendan F. Klare et al., "Face Recognition Performance: Role of Demographic Information," *IEEE Transaction on Information Forensics and Security* 7, no. 6 (December 2012): 1789, cited in Clare Garvie, Alvaro M. Bedoya, and Jonathan Frankle, "The Perpetual Line-Up: Unregulated Police Face Recognition in America," Georgetown Law Center on Privacy & Technology, October 18, 2016, https://www.perpetuallineup.org/.

that 80 percent of the tens of millions of Americans: Olivia Solon, "Facial Recognition Database Used by FBI Is Out of Control, House Committee Hears," *Guardian,* March 27, 2017, https://www.theguardian.com/technology/2017/mar/27/us-facial-recognition-database-fbi-drivers-licenses-passports.

An in-depth study by ProPublica found numerous cases: Julia Angwin et al., "Machine Bias," ProPublica, May 23, 2016, https://www.propublica.org/arti cle/machine-bias-risk-assessments-in-criminal-sentencing.

Biased outputs have been detected: Executive Office of the President, *Big Data: Seizing Opportunities, Preserving Values* (Washington, DC: White House, May 2014), 58–59, https://obamawhitehouse.archives.gov/sites/default /files/docs/big_data_privacy_report_may_1_2014.pdf.

a group of RAND researchers found: Osonde A. Osoba and William Welser IV, *An Intelligence in Our Image: The Risks of Bias and Errors in Artificial Intelligence* (Santa Monica, CA: RAND Corporation, 2017), 3.

14. Rules for the Eye and the AI

214 *When the Department of Homeland Security:* Jay Stanley, "A Company Announces Its Intent to Create World's Newest Privacy Nightmare," ACLU Blog, April 23, 2018, https://www.aclu.org/blog/privacy-technology/surveil lance-technologies/company-announces-its-intent-create-worlds-newest.

215 *Baltimore's memo outlining the legal basis:* Baltimore Police Department, "Memorandum of Law in Support of Constitutionality of Wide Airborne Surveillance."

there are over 600 police departments: Dan Gettinger, "Public Safety Drones: An Update," Center for the Study of the Drone, May 28, 2018, https://dronecenter.bard.edu/public-safety-drones-update/.

a set of vague voluntary best practices: US National Telecommunications and Information Administration, *Voluntary Best Practices for UAS Privacy, Transparency, and Accountability*, May 18, 2016, https://www.ntia.doc.gov/files/ntia/publications/uas_privacy_best_practices_6-21-16.pdf.

216 *Take the case of Santiago, Chile:* Derechos Digitales, "Lo que la Corte Suprema no comprende sobre los globos de televigilancia," June 8, 2016, https://www.derechosdigitales.org/10051/lo-que-la-corte-suprema-no-comprende-sobre-los-globos-de-televigilancia/.

the country's constitution: Comparative Constitutions Project, *Chile's Constitution of 1980 with Amendments Through 2015*, https://www.constitutepro ject.org/constitution/Chile_2015.pdf?lang=en, p. 9.

"It is clear that people living under the balloons": Derechos Digitales, "Lo que la Corte Suprema no comprende sobre los globos de televigilancia."

The court drafted a list of conditions of use for the city government: Ibid.

217 *The mayor of one of the neighborhoods even hired:* Pino M. Patricio, "Acusan que globos de vigilancia en la zona oriente son 'orwellianos,'" *La Segunda*, August 17, 2015, impresa.lasegunda.com/2015/08/17/A/FU2OHSKE/all.

In the 1930s, soon after lawmakers clamped down: Meyer Berger, "Tapping the Wires," *New Yorker*, June 18, 1938, https://www.newyorker.com/magazine/1938/06/18/tapping-the-wires.

"a system that uses one or more": Rahul Thakkar, ed., *A Primer for Dissemination Services for Wide Area Motion Imagery*, OGC 12-077r1 (Wayland, MA: Open Geospatial Consortium, 2012), 7.

218 *Congress didn't debate a serious bipartisan:* Erin Kelly, "Bipartisan Bill Seeks Warrants for Police Use of 'Stingray' Cell Trackers," *USA Today*, February 15, 2017, https://www.usatoday.com/story/news/politics/onpolitics/2017/02/15/bipartisan-bill-seeks-warrants-police-use-stingray-cell-trackers/979 54214/.

219 *the FBI did not clarify whether it requires agencies:* Ellen Nakashima, "FBI Clarifies Rules on Secretive Cellphone-Tracking Devices," *Washington Post*,

May 14, 2015, https://www.washingtonpost.com/world/national-security/fbi-clarifies-rules-on-secretive-cellphone-tracking-devices/2015/05/14/655b4696-f914-11e4-a13c-193b1241d51a_story.html?utm_term=.552514416d93.

the NYPD embarked on an aggressive: Adam Goldman and Matt Apuzzo, "With Cameras, Informants, NYPD Eyed Mosques," Associated Press, February 23, 2012, https://www.ap.org/ap-in-the-news/2012/with-cameras-infor mants-nypd-eyed-mosques.

220 *in 2016 the Brazilian government:* Juanma Rubio, "Brasil invierte 7,1 millones en cuatro globos de seguridad. Esta tecnología salta al mundo civil un sistema utilizado en Irak y Afganistán," AS.com, August 1, 2016, https://as.com/masdeporte/2016/08/01/juegosolimpicos/1470016907_654948.html.

Because of an anticorruption law: John Marion, email to the author, March 8, 2017, and interview with the author, February 10, 2017.

Two of the four blimps were powered by generators . . . 600 feet: "Balões de Vigilância — Jogos Olímpicos Rio 2016," *Konatus,* March 2, 2017, http://www.konatus.com.br/baloes-de-vigilancia-jogos-olimpicos-rio-2016/.

221 *according to Logos's John Marion, the analysts spent much of their time:* Marion, interview, February 10, 2017.

In 2004 NYPD officers monitoring a protest: Kim Zetter, "What an NYPD Spy Copter Reveals About the FBI's Spy Planes," *Wired,* June 5, 2015, https://www.wired.com/2015/06/fbi-not-alone-in-operating-secret-spycraft/.

222 *In two rulings in 2010 and 2012:* Orin Kerr, "DC Circuit Introduces 'Mosaic Theory' of Fourth Amendment, Holds GPS Monitoring a Fourth Amendment Search," Volokh Conspiracy, August 6, 2010, volokh.com/2010/08/06/d-c-circuit-introduces-mosaic-theory-of-fourth-amendment-holds-gps-monitoring-a-fourth-amendment-search/.

More recently, in Carpenter: Carpenter v. United States, No. 16-402, 585 U.S. (2018).

223 *The study, from 2015, is primarily about drones:* Richard M. Thompson II, *Domestic Drones and Privacy: A Primer* (Washington, DC: Congressional Research Service, March 30, 2015), 6.

224 *Steve Suddarth suggested:* Steve Suddarth, interview with the author, June 9, 2017.

225 *Google already does exactly this:* Dave Barth, "The Bright Side of Sitting in Traffic: Crowdsourcing Road Congestion Data," Google Official Blog, August 25, 2009, https://googleblog.blogspot.com/2009/08/bright-side-of-sitting-in-traffic.html.

226 *an investigation by the* Los Angeles Times: Jazmine Ulloa, "California Passed a Law Boosting Police Transparency on Cellphone Surveillance. Here's Why It's Not Working," *Los Angeles Times,* August 27, 2017, www.latimes.com/politics/la-pol-sac-cell-phone-surveillance-transparency-law-20170827-html story.html.

Similarly, as of late 2017 privacy groups: Pablo Viollier, Public Policy Analyst, Derechos Digitales, interview with the author, July 13, 2017.

229 *says Mr. Weasley:* Greg Zacharias, "Air Force Future Operating Concept: Implications for Autonomy," PowerPoint presentation, Mitchell Institute, Portland, ME, May 17, 2016.

DARPA's Explainable AI program: David Gunning, "Explainable Artificial Intelligence (XAI)," PowerPoint presentation, Defense Advanced Research Projects Agency, November 2017.

230 *When I called Chuck Blanchard:* Charles Blanchard, interview with the author, July 26, 2017.

231 *So much so that even companies like Microsoft:* Brad Smith, "Facial Recognition Technology: The Need for Public Regulation and Corporate Responsibility," Official Microsoft Blog, July 13, 2018, https://blogs.microsoft.com/on-the-issues/2018/07/13/facial-recognition-technology-the-need-for-public-regulation-and-corporate-responsibility/.

232 *A study by a White House task force:* Executive Office of the President, *Big Data: Seizing Opportunities, Preserving Values* (Washington, DC: White House, May 2014), https://obamawhitehouse.archives.gov/sites/default/files/docs/big_data_privacy_report_may_1_2014.pdf, p. 5.

Dozens of towns and counties: D. J. Pangburn, "Berkeley Mayor: We Passed the 'Strongest' Police Surveillance Law," *Fast Company,* April 24, 2018, https://www.fastcompany.com/40558647/berkeley-mayor-we-passed-the-strongest-police-surveillance-law.

In 2017 New York City adopted a bill: Roshan Abraham, "New York City Passes Bill to Study Biases in Algorithms Used by the City," Motherboard, December 19, 2017, https://motherboard.vice.com/en_us/article/xw4xdw/new-york-city-algorithmic-bias-bill-law.

Index